What Kind of Ancestor Do You Want to Be?

What Kind of Ancestor Do You Want to Be?

Edited by John Hausdoerffer,
Brooke Parry Hecht, Melissa K. Nelson,
and Katherine Kassouf Cummings

The University of Chicago Press   CHICAGO & LONDON

The University of Chicago Press, Chicago 60637
The University of Chicago Press, Ltd., London
© 2021 by The University of Chicago
All rights reserved. No part of this book may be used or reproduced in any manner whatsoever without written permission, except in the case of brief quotations in critical articles and reviews. For more information, contact the University of Chicago Press, 1427 E. 60th St., Chicago, IL 60637.
Published 2021
Printed in the United States of America

30 29 28 27 26 25 24 23 22 21   1 2 3 4 5

ISBN-13: 978-0-226-77726-9 (cloth)
ISBN-13: 978-0-226-77743-6 (paper)
ISBN-13: 978-0-226-77757-3 (e-book)
DOI: https://doi.org/10.7208/chicago/9780226777573.001.0001

Library of Congress Cataloging-in-Publication Data

Names: Hausdoerffer, John, editor. | Hecht, Brooke Parry, editor. | Nelson, Melissa K., editor. | Cummings, Katherine Kassouf, editor.
Title: What kind of ancestor do you want to be? / edited by John Hausdoerffer, Brooke Parry Hecht, Melissa K. Nelson, and Katherine Kassouf Cummings.
Description: Chicago : University of Chicago Press, 2021. | Includes bibliographical references and index.
Identifiers: LCCN 2020043082 | ISBN 9780226777269 (cloth) | ISBN 9780226777436 (paperback) | ISBN 9780226777573 (ebook)
Subjects: LCSH: Traditional ecological knowledge. | Indians of North America—Social life and customs.
Classification: LCC GN476.7 .W438 2021 | DDC 970.004/97—dc23
LC record available at https://lccn.loc.gov/2020043082

For flow and clarity, some interviews have been edited.

♾ This paper meets the requirements of ANSI/NISO Z39.48-1992 (Permanence of Paper).

CONTENTS

Introduction 1
POEM: Unsigned Letter to a Human in the 21st Century, *Jamaal May* 6

I. EMBEDDED
*Our ancestral responsibility is deeply rooted in
a multigenerational relationship to place.*

A. POEM: Great Granddaddy, *Taiyon Coleman* 13
B. ESSAYS:
    I. Ancestor of Fire, *Aaron A. Abeyta* 16
    II. Grounded, *Aubrey Streit Krug* 20
    III. My Home / It's Called the Darkest Wild, *Sean Prentiss* 25
C. INTERVIEW: Wendell Berry, *Leah Bayens* 33
D. POEM: To the Children of the 21st Century, *Frances H. Kakugawa* 44

II. RECKONING
*Reckoning with ancestors causing and
ancestors enduring historical trauma.*

A. POEM: Forgiveness?, *Shannon Gibney* 49
B. ESSAYS:
    I. Sister's Stories, *Eryn Wise* 51
    II. Of Land and Legacy, *Lindsey Lunsford* 56
    III. Cheddar Man, *Brooke Williams* 61
    IV. Formidable, *Kathleen Dean Moore* 68

C. INTERVIEW: Caleen Sisk, *Brooke Parry Hecht and Christopher (Toby) McLeod* 73
D. POEM: Promises, Promises, *Frances H. Kakugawa* 84

III. HEALING
Enhancing some ancestral cycles while breaking others.

A. POEM: To Future Kin, *Brian Calvert* 89
B. ESSAYS:
   I. Moving with the Rhythm of Life, *Katherine Kassouf Cummings* 90
   II. (A Korowai) For When You Are Lost, *Manea Sweeney* 95
   III. To Hope of Becoming Ancestors, *Princess Daazhraii Johnson and Julianne Lutz Warren* 100
C. INTERVIEW: Camille T. Dungy and Crystal Williams 110
D. POEM: Yes I Will, *Frances H. Kakugawa* 117

IV. INTERWOVEN
Our descendants will know the kind of ancestor we are by reading the lands and waters where we lived.

A. POEM: Alive in This Century, *Kristi Leora Gansworth* 121
B. ESSAYS:
   I. What Is Your Rice?, *John Hausdoerffer* 124
   II. Restoring Indigenous Mindfulness within the Commons of Human Consciousness, *Jack Loeffler* 130
   III. Reading Records with Estella Leopold, *Curt Meine* 138
   IV. How to Be Better Ancestors, *Winona LaDuke* 142
C. INTERVIEW: Wes Jackson, *John Hausdoerffer and Julianne Lutz Warren* 145
D. POEM: Omoiyare, *Frances H. Kakugawa* 154

V. EARTHLY
Other-than-human beings are our ancestors, too.

A. POEM: LEAF, *Elizabeth Carothers Herron* 157
B. ESSAYS:
   I. The City Bleeds Out (Reflections on Lake Michigan), *Gavin Van Horn* 159
   II. I Want the Earth to Know Me as a Friend, *Enrique Salmón* 166
   III. The Apple Tree, *Peter Forbes* 172
   IV. Humus, *Catriona Sandilands* 177
   V. Building Good Soil, *Robin Wall Kimmerer* 182

C. INTERVIEW: Vandana Shiva, *John Hausdoerffer*   185
D. POEM: Your Inheritance, *Frances H. Kakugawa*   194

VI. SEVENTH FIRE

A. POEM: Time Traveler, *Lyla June Johnston*   197
B. ESSAYS:
  I. Seeds, *Native Youth Guardians of the Waters 2017 Participants and Nicola Wagenberg*   200
  II. Onëö' (Word for Corn in Seneca), *Kaylena Bray*   210
  III. Landing, *Oscar Guttierez*   214
  IV. Regenerative, *Melissa K. Nelson*   219
  V. Nourishing, *Rowen White*   225
  VI. Light, *Rachel Wolfgramm and Chellie Spiller*   232
C. INTERVIEW: Ilarion Merculieff, *Brooke Parry Hecht*   240
D. POEM: Lost in the Milky Way, *Linda Hogan*   252

Acknowledgments 255
Notes 257
About the Contributors 263
Index 273

INTRODUCTION

Morning on the coast brought the kind of weather one expects from spring at the ocean's edge. Northwest of San Francisco, heavy fog lifted slowly, revealing verdant hills upon which dairy cows and tule elk were feasting. The dewy chill settled around us as we hastened to the ceremony. As we approached, we could smell burning sage. We gathered in a circle and listened as Coast Miwok leader Sky Road Webb offered his gratitude to the ancestors who shaped his coastal homeland here on the land taken and called California. Cracking open an oyster shell, he laid the burning sage inside, asking each of us to hold the shell so we could invite our own ancestors into the circle. Doing so, we smudged ourselves with the burning sage. The shell, the sage, the community, and Webb all merged bringing together sea, land, sky, people, past, present, and future to reflect on the question that gathered us that day, beginning our collaboration as coeditors: *What Kind of Ancestor Do You Want to Be?*

One of our coeditors first heard this question while interviewing Winona LaDuke on the White Earth Reservation of Minnesota. As LaDuke prepared her daily dinner for her household of children, neighbors, interns, staff, and activists, Anishinaabe elder Michael Dahl posed the question: *What Kind of Ancestor Do You Want to Be?* We view this compelling question as *eternally urgent*. *Eternal* because it calls forth ancient wisdom and multigenerational ethics necessary for any human community to survive and thrive. *Urgent* because the planetary impacts of colonial overconsumption of resources and domination of peoples dramatically threatens the livability of this planet.

More than asking us how we want to be remembered, the question of *What Kind of Ancestor Do You Want to Be?* suggests that we are, always and already, ancestors—even if we never are remembered or never have children. The question deepens our awareness of the roots and reach of all of our actions and non-actions. In every moment, whether we like it or not and whether we know it or not, we are advancing values and influencing systems that will continue long past our lifetimes. These values and systems shape communities and lives that we will never see. The ways we live create and reinforce the foundation of life for future generations. We are responsible for how we write our values, what storylines we further and set forth—the world we choose to cultivate for the lives that follow ours. So how are we to live?

The role of ancestor relies upon the realization of our descendants. Ancestry is an intergenerational practice of becoming, of change, in a world that, Robin Wall Kimmerer reminds us, "is constantly in flux." Just as we prepare the young to step into adulthood and release childish ways for the health and growth of society, so the practice of becoming an ancestor requires the release of our grip upon what is, the letting go of certain ways of being in the world to embrace the changes required for the stream of life to keep flowing vigorously. We can begin to be living ancestors by stepping aside and handing over our power, our place, and our voice in ways that nurture those coming up. Those may be your own child, your grandchild, your neighbor's child; those may be your former student, your young friend, your young colleague; those may be the young of other species, as Kimmerer writes in this volume, "the grandchildren of warblers, bumblebees, and hemlocks." Ancestors empower the living. A good ancestor empowers the living to live well, so that life may flourish and the channel between ancestors and descendants may continue to flow. Being a good ancestor means understanding how to handle power, when to hold it, when to hand it over, and how to transform it. Just as our bodies, the fundamental sources of our power, will be transformed when breathing ends, so we can begin to practice being ancestors today by recognizing how we want to transform the power—the life force—we've been given. As Catriona Sandilands writes in this volume, "Good ancestry is . . . a generous offering of one's life to an unknown, multiplicitous future."

What Kind of Ancestor Do You Want to Be? is a question that we, the editors, believe is more important than ever to ask in 2020—when our communities have become fragmented by a global consumer society, when our selves have become isolated in a competitive and technologically driven economy, and when our spiritual, social, and ecological impacts on human and other-than-human beings extend farther than ever imagined due to globalization and climate change. In the midst of an ever-increasing upending of life, it is vital to reimagine how we live. We believe that this is a question that catalyzes dreams—dreams rooted in the instructions of the peoples and places that came before us; dreams that stretch our imagination long beyond our time on this sacred planet.

While we believe this question holds universal relevance for all members of our global community, it is rooted in the ethical traditions of the diverse Indigenous communities of Turtle Island. One of our coeditors, Melissa Nelson, from the Anishinaabe and Métis traditions, speaks to these roots:

> As Indigenous communities, as tribes, as place-based citizens, as human beings, we know we would not exist without the love and struggle of our ancestors. Our DNA reminds us of this bond, what the late Dakota poet John Trudell called our "Descendants-N-Ancestors." Because of this, through the links of time and countless generations, we strive to honor and recognize our ancestors in ceremony and ritual, song and dance, and daily acts of gratitude. "We are all related," even to the unseen, and we recognized that our ancestors are with us in the "river of souls" (Milky Way), in the corn that we grow and eat, in the water we splash on our face every morning.
>
> "We are the result of the love of thousands," Linda Hogan reminds us. We exist because of the love and instinctual survival of our ancestors, and our love and regenerative power will create future generations, all those to come who are emerging from the ground of being to replace us as we return to Earth.
>
> We are like trees with deep roots connected to our origins in the Earth, in soil, in rock, in stardust. We come from strong trunks, and numerous branches of ancestors loving and living and recreating to make our particular branch and ancestral thread. These are our family trees, our genealogies, what our Maori relatives call their *whakapapa*. In reflecting on what kind of ancestor we want to be, we wonder, how does the outer needle of a redwood branch 250 feet in the sky remember its ancestors in the redwood roots?
>
> We are like rivers flowing from humble springs, where freshwaters emerge from primal wombs of fertility and rebirth. We are the result of all those ancestral waters flowing back to the liquid Mother. As we travel down mountains and along riverbanks, flowing between roots amongst liquid minerals and sediments, we merge with all those stories, all those memories, all those life forces. As our waters flow and grow and transform, we make habitat for new life—for spawning salmon, for mating otters, for sparring dragonflies, and other animal delights. This procreativity gives new life to new generations. We wonder, does a splash of freshwater meeting the salty sea for the first time acknowledge its ancestors in the mountain spring, in the river fog, in the thick, gray clouds? Ancestral waters are explored and honored throughout this book.

We are like seeds holding the sacred spark of life, ready for new growth when the conditions are right. We come from the kernels of our parents and carry their DNA within us. Throughout our lives we carry that fertility and potentiality for regeneration. That life force has many names to our people. It is the mysterious spirit of life itself, the *manitou* for Anishinaabeg; it is *mana* for the Hawaiians and Maori. It has many names even though it is ultimately nameless. Many young people, you will see, are exploring their ancestors through renewed relationships with their ancestral seeds and food traditions.

We are like fire, a spark of lightning meeting a pine snag and then two become one fire that consumes and grows with fuel and air. Fire is our ally, although we have made it an enemy. Winnemem Wintu Tribal Chief Caleen Sisk reminds us of the sacred role of fire as a revered teacher to Native peoples. Humans are currently out of balance with fire, thus we are experiencing devastating fires as we have forgotten that ancestral connection and need to be in right relation with it as an ancestor and relative. Noptipom Wintu ethnobotanist Sage LaPena reminds us that our relationship with outer fire recalls our relationship to our inner fire, and that too has to be tended to rebalance.

According to prophecy, twenty-first-century Anishinaabeg (Ojibwe, Potawatomie, and Odawa Native Peoples on Turtle Island) are people of the Seventh Fire, in a time of remembering and repairing and revitalizing ancestral lifeways that were disrupted from five hundred years of colonization. They are preparing and hoping to light the Eighth Fire, a time of a new people committed to peace and justice. To light this new, and prophecy says *last*, fire, they look to their human ancestors, especially the youth, and more-than-human ancestors, to cosmogenealogies that remind them we are all related, to the trees, soils, stars, seeds, and sacred sources of water and fire.

Please join us in asking this question of yourself, your family, your community, your human and more-than-human relations and elders, your ancestors and your descendants. We hope we have devised an engaging and enjoyable way for you to explore the question.

First, we have divided the book into six parts. "Embedded" explores the way in which being deeply connected to a place leads to understanding that our most powerful ancestors are those whose work, livelihood, hopes, and vision have shaped the environment. "Reckoning" investigates ways in which we struggle with past injustices to enact a more just future, whether through rethinking the violence that one's position of privilege enjoys or understanding the exploitation exacted upon one's own people long ago. "Healing" presents ways of recovery from cultural traumas faced by ancestors, enabling future generations to thrive. "Interwoven" looks to expansive ways in which we, as ancestors, will be remembered by how our values relate to the land and living communities. The integrity of our ancestry, ultimately, is predicated upon the integrity of the beings, systems, and communities that our values and practices enhance. "Earthly" looks for ancestors in the other-than-human world, and for instructions on how to live as humans from our other-than-human ancestors—other species and ecological communities. "Seventh Fire" returns us to how this journey began. Just as this book began with a question from Michael Dahl on Winona LaDuke's White Earth Reservation back deck, so this book will look to the next generation of Indigenous leaders for the questions to guide us all as ancestors-in-practice. After pondering this question of eternal urgency, their messages are the ones we hope will remain in your hearts and minds and that you may live out into the world.

Finally, a word on the format. We hope you enjoy the mix of genres and the rhythm of forms in this book. We believe that the *Ancestor* question indeed taps into something ancient, something that the poets and oral storytellers of diverse cultures have talked about since the beginning of speech. Each section opens with a poem, continues with essays from varied perspectives and diverse authors on the question, moves into a conversational interview with a leading figure who has considered this question in the way they live, and closes with a poem. We hope this pattern will be enjoyable for all readers, and allow for you to settle into each section's theme in one sitting. We also hope the mix awakens your senses, opens your heart, and engages you in dialog as you contemplate the many essays from numerous cultural communities—as you begin to respond to the *Ancestor* question yourself.

We are deeply grateful that you have joined us on this journey. You are our most important ancestors, those who will contemplate this question and renew its relevance long into the futures of humans and of all of our relations.

In Gratitude,
John, Brooke, Melissa, and Katherine

Poem: Unsigned Letter to a Human in the 21st Century

Jamaal May

Dear citizen of the binary mirror,
Dear wide-eyed and deft-fingered,
Dear deer in an ex-forest clearing,

Please forgive me if I'm speaking too soon.
Are you here yet?
Or is it that your presence only looks like arrival
because you can't help but be the loudest ghost?
If that wasn't you I saw
moving toward the edge of a cliff
like a moth to a moth
that flutters in flames,
no worries, I'll leave this here for another time.

I write you because you asked me to,
though you may not remember the call,
though we may not have ever spoken.
But I heard it in your verse
after you bit your tongue
and bragged about the blood.
Dear aspirant to the throne
of Most Unassailable Victim,

I write you because when I opened my mouth
to say *love*, someone said sword again.

But Rumi said *love*,
and so I'm out in New England again
and in Detroit again
And in America still
with my ear pressed to this red book
and sure enough, the harsh sound
of a scabbard emptying
so that a belly could fill with blade
was drowned out for a moment,
as Love went running down the mountainside
out into the everything, becoming
the mad man again.
I closed my eyes and opened
what I didn't know was locked inside,
and even now, I fumble for its name.
Could it have been the opening itself?

I write to confess I've known for years
that the closed-off parts of me
need the closed-off parts of you.
You are full of more you than is known,
and I confess I never bothered to notice.
But now I notice everything.
How my sink holds a sea,
overflows and becomes a
waterfall, becomes a puddle
at my feet. There's an ancient article
from just seven years ago on waterboarding.
Another provides a litany
of the best ways to be ok with everything
our hearts stutter about: the bleach-white coral,
the pistol-full avenues,
the lucrative penitentiaries,
and women's shelters that bulge with bruise.
They say the starved are guilty
of being hungry, and the head-scarfed girl
is guilty of living near the detonation,
and now I notice my hands—
how little they hold. They are teacups

bailing out this vessel
not worthy of any sea, and silence
is the rag stuffed into the mouth
of a gurgling drain. But I digress.

Dear digital city,

Scrolling around our world's web
I opened another window
full of the seething news.
But this time, I could only laugh
at what Destruction had cloaked itself in.
If the gauze-like veil had a name, it would be
something along the lines of *Inevitability*.
But you are training in an alchemy
that can make those first three syllables fall away.
Destruction doesn't know the strike
in the middle of it is past tense;
let the shark spin in its cage.
We only need to live long enough to teach
those who will, tomorrow,
drown it in the air.

My friend,

I write because I love you enough
to ask for what is terrible: run farther
than your feet can possibly carry your heart.
I love you enough to confess that you will fail
but fail closer to the finish line
than if you lie down when the start gun fires.
And in this way, you will never fail
to be arch, stepping-stone, bridge
of bone and intellect,
of guts and song. Look
how lively the children step.
Let's nod our heads to their footfalls.
Become backbeat with me
and they will sing the harmonics
we forgot to learn.

Tell me you wouldn't die for that.

Tell me you will live for this.

Love,

[Reprinted with the permission of Alice James Books, from *The Big Book of Exit Strategies*]

1

Embedded

I am asked what kind of ancestor will I be. I struggle with the answer because I mistake it for asking about the future, but really it is calling me back to the places where I learned how to be.

AARON A. ABEYTA

POEM

Great Granddaddy

Taiyon Coleman

(for Frank Coleman)

1925 is a long time to be born black
outside of Greenwood, Mississippi,
the Choctaw and Chickasaw Delta
whose rich tributaries spit cotton
and soil and grow the grass
that your father's only horse grazes
after he rides him in hidden places
between honey bees and leaves
while wearing a black top
hat that makes the white men,
who hide and sneak
in between the tall trees,
to buy and drink
your father's homemade
whiskey, mad.

1940 gives its first taste of spring
and you are excited. Your father
has taught you how to search
and pick the right plants and herbs
for certain people,
in certain places,
and on certain occasions,
and today, your father

lies in his bed,
too sick to stand,
too sick to move,
and you are proud
that he trusts you
to gather herbs
to soothe
and to cure
what ails him.

Lavender and blue flowers
and wet spring roots are bright
in your black hands when you return
to the cabin in the Mississippi
woods, hesitant at first to enter
because the one horse that was
there when you first
left is now gone,
and your father's black top
hat never far from his dusky
bald head, lies broken
and smashed in the red
clay dirt.

Intent in your task
as you are a good son,
and you know that children
are meant to be seen and not
to be heard, especially black
children, you boil the right
spring flowers and the right
spring roots while praying
to the right Gods,
honoring the right ancestors,
and sacrificing to the right
vengeful Spirits, you let
it cool for the right
amount of time, you drain
it through cleaned cheesecloth,
and you spoon it directly

to your father's wide mouth
while holding his head
in your left hand
until you trust
that it's enough.

Because he says nothing,
you know that you have
done a good job,
and three days later,
you will still
bury him in the secret
place where he told you
to cover him if a thing
like this happens,
and you conceal
the dirt and leaves
that conceal him
with your left-over
blue and purple flowers,
and orange roots.

And your light-skinned
momma, my great grandma
on her way to St. Louis,
will speak of forbearers,
of predecessors,
of poison,
and of a black man daring
to ride a brown spotted horse
while wearing a shiny black
top hat, missing jugs
of whiskey and money
to last a season,
and how easy
it is to kill
when sharing food
and water with someone
you trust.

ESSAY

Ancestor of Fire

Aaron A. Abeyta

I wish to be an ancestor of fire and how flames came to be, an ancestor of particular slants and dances of light. We are working in the half-light of a March dusk; the lambs run and jump as if they are born of sun reflected on ice. They rush to their mothers and their tails work furiously with joy as they suckle. Even then, as a boy, I somehow understood that those lambs, in that light, stood for hope and something greater than joy.

A few hundred yards to the south there are intermittent cracks of metal on wood, the unmistakable tone of my grandmother chopping wood. The small adobe house is at her back and the smoke of pine and aspen lingers above her home. My abuelita will take her load of wood, split thin as her wrists and the length of her forearm. She will stoke the fire and, on the steel stovetop that shines of countless fires, she will place milk to warm in a large pot.

At this distance, above the cacophony of dusk and putting the animals under the shed for the evening, I can hear my grandmother's orphan lambs crying for their supper. She has them in a small pen next to the chicken coop and I am able to imagine them pressed against the chicken wire, their bleating a siren of hunger and of love for the only mother any of them have ever known. Every year she raises at least a dozen lambs. The circumstances of their orphandom are consistent, there is little in the way of variation—a mother ewe with too little milk to raise twins or triplets; a mother ewe too old; a mother ewe dead of bloating or a prolapsed uterus gone septic; or sometimes a mother ewe that simply did not know how to be a mother; and then there are the lambs that did not suckle, were sat upon by their oblivious mothers or those born out in the open, without shelter, nearly frozen to death, the ones we bring to the wood stove in the kitchen, the lambs we lay on old towels and blankets and feed with a dropper or a syringe, the ones that rarely survive but whose life we honor with our attempt at saving it. By the end of lambing season there are at least a dozen, usually more, that she has taken in. To the uninitiated they seem indistinct and impossible to tell apart. She knows them by sight, knows which she has fed and which she has not; she has given them simple and obvious names, the lamb with the spots on her nose is Pinto; the all white lamb is Blanca; the energetic one is Jumper and the smallest is Tiny; there is always a lamb named Curly and one named Brownie.

Every dawn, again at noon and then again at dusk, she makes her way to the pen next to the chicken coop; they hear her approach and they crowd the gate, poking their noses through the wire. Each lamb has its own bottle, but feeding them is chaos, they always seem so desperate and hungry, as though they may never eat again, the lambs always pressing at the bottle and feeding of the others; they are never full. Hunger seems to be the only memory the lambs possess, that and who their mother is, this short and stoic woman who carries two buckets filled with 12-ounce bottles to them three times a day. They bleat and sing out to her, the only mother they know or remember. We call them, collectively, *pencos*. My grandmother is blessed with the grace of loving things that others cannot or will not; living things are her gift.

On Fridays she bakes bread and every Sunday she roasts a chicken; she seems made entirely of work, custom, and survival. This is the ancestor I will be, like my grandmother working in the evening light. Did she know back then that she was teaching me to care for fallen and forgotten things? I remember her most often at the woodpile. She works in the wood chips and dust, generations of split wood, looking for slivers of pine and aspen, discarded reeds she knows will take a flame from a single match. She collects what others do not notice. She places her finds in a tin bucket. She saves what no one else uses or needs. Her cupboards are filled with homemade jelly and jars of buttons. Her actions tell me to collect everything; someday it will be of use. She precedes me, my ancestor.

I remember her working in the woodpile at dusk. One never imagines such a flame as exhaustible. Youth conspires to render such things as permanent. That fire is love and that love is fire, and neither can be extinguished. Age, however, puts truth into everything; it is the tree as it falls. Truth is the enemy of memory. Truth is the well from which brokenness is pulled up in a tin bucket. Were it not for truth she might be baking bread for centuries, collecting sticks that will always take a flame and mending shirts, every missing button easily replaced. I am a grandchild of a woman who knew work as innate, necessary, a direct descendent of survival. My abuelita's faith was made not of prayer but of hope in collected things, rituals of kindness and patience.

I am told that I am an old spirit. I wonder sometimes how many generations I carry in me. Odd thoughts follow me like orphan lambs in single file behind their human mother. I wonder why I love the smell of smoke; why well water is sweetest when sipped; why lambs at their end, sprawled in front of the wood stove, thrash their legs as if running. Is every jar of jelly so sweet as to be undefinable? Must bread be baked only on Fridays? Will a bucket filled with sticks light enough fires to last for centuries?

Her house is empty now, barely standing but very much standing. The red shingles on the roof are in need of repair, the pictures, the kitchen table, the rooms are all like monuments to their last inhabitants. Her wood stove was removed, now in the garage; the wood box is empty, shelves barren as a January death. Silence lives there now, where before there was so much noise. Everything seems to be falling or fallen, the pens and the chicken coop are gone, the ditch that cut the yard in two rarely runs now; her lawn and her garden are likely gone forever. The woodpile is gone too, only the slivers of dust and splinters of wood scattered and disintegrating. My ancestor walked here once. I can see her move slowly from the front door and into her yard. There is a picture of her at her front gate; the picture is grainy, her glasses are dark and her eyes invisible behind them. She looks to the south, toward the camera; she is wearing a pink sweater, unbuttoned, over her white house dress. Her shoes are white, nurses shoes, the only kind she wore. The meadow is behind her, and the field is new green and the leaves of the distant cottonwoods are beginning to fill in. My abuelito's blue Ford is parked to her left. Down the hill from her there is an old wagon, falling apart, a testament to the fact that everything ends in silence. On the sidewalk, on the other side of the gate, there are four lambs, big and weaned, looking up to their human mother, whose left elbow rests upon the closed gate. The picture lacks focus, but she is almost smiling; the corral is in the distance, beyond it the river, beyond the river there are trees and beyond the trees there is a mountain and beyond the mountain there is time, and there is nothing beyond time, memory tries to reach that far but cannot. Memory outlasts flesh and flesh outlasts grief, and grief is the natural heir of loss, and loss is the acknowledgment of love diminished and love diminished is the ancestor of love in full, and love in full is the ancestor of time. Her smile was a rare gift. I once compared it to an eclipse. I will be an ancestor of fire. That is what I was always meant to be.

The orphan lambs call into the last of the light, so she pours warm milk into 7-Up bottles, tops each of them with a black rubber nipple. She moves slowly.

The green bottles clank and chant in yet another tin bucket. She carries more than just a meal to her lambs. They love her and their tails prove it is so. The fire in the kitchen burns. The lambs sing their bleats and baaahs into the dusk. She feeds them and they become full. It has been this way forever; only breath is older.

Where there is absence, bring to that place warm milk and love. Where others discard, there is the work of living and saving.

I am asked what kind of ancestor will I be. I struggle with the answer because I mistake it for asking about the future, but really it is calling me back to the places where I learned how to be. In the distance of back then, the light is gone from the day. Everything has folded into purple and black. I imagine a single light; it glows in the vastness of memory. I hear wood split. Smoke ascends. I am a boy, again. I must have been helping the men lock up sheep for the evening, but just beyond the corral my ancestor is chopping wood, preparing for the night. The wood box is full. Seek, she tells me; collect everything; care for lost things; she compares love to bread baked in a wood-burning stove. She has gone before me. This flame is sacred. I am an ancestor of fire. Time has no limits.

ESSAY

Grounded

Aubrey Streit Krug

I started to feel the absence of Bluestem roots around me when I was sitting in a basement classroom at the land-grant University of Nebraska. In that prairie place a few centuries ago a community of soil and tallgrass was tended by peoples and creatures, fed by sunlight and rain and rock. Then they were removed.

The human newcomers who broke ground there might have thought they were putting the past behind them, eventually making the land into a "learning environment" for students like me. I grew up in a small wheat-farming town in nearby Kansas. The commonsense agricultural assumptions I inherited were so normal and natural to me as to be invisible. Only later did I begin to see with a shock these inherited ideas about life as a crop growing upward and forward toward a future harvest; ideas about settling fields and extracting value and digging up the roots of any problems that stand in the way. I am in part the result of ancestral soil communities being unearthed as if they were already dead, as if they were never alive.

Then I sat in that basement classroom, and the Bluestem roots began weighing on my shoulders and tangling through my hair. Their absence became physically real to me for the first time. Feeling them was distinctly, well, unsettling. They exerted the pressure of memory. I felt the constraints of my lens upon reality begin to soften, along with the edges of my human self, into humus. I was troubled. Was this being wounded? Was this healing?

*

The grounds of our planet's being have come about in part because of the movement of distant generations of plants. Biologists Linda E. Graham,

Martha E. Cook, and James S. Busse explain, "The origin of terrestrial ecosystems, including the development of modern soils and their biota and emergence of terrestrial metazoan groups, depended on the previous colonization of land by the ancestors of modern land plants."[1]

In the deep and long-standing perspective of plants, whose sensory awareness is perhaps one way the land knows us, humans evolved from tree-dwellers.[2] Only a short while ago we moved out into the grasslands and savannas. Then, just recently, we started domesticating and cultivating the seeds of grasses—rice, wheat, corn. In the perspective of plants, human history is young, and yet we've washed away tons of topsoils in an instant.

To me, it feels like just yesterday that the Dust Bowl blew here. Here, this pasture, in the hills near my hometown, with blades of Little Bluestem sparking green in the limestone. My grandma married my grandpa in Kansas the year that drought began. Droughts belong on these Great Plains, but from what I understand of climate models for the rest of this century, there is the chance for coming droughts and dust storms to be greater than any farmers in my family line have felt. In the pasture, watching my dad and little boy walk together, I envision the questions of weathered children. Will they look at me, at all of us, and say, "How *could* you, knowing?" And will I stand in front of them unable to say a word?

*

Several years ago, back when I was at the University of Nebraska, I became a student of Umóⁿhoⁿ íe tʰe, the Omaha language. Learning has not been an easy process, and I am far from fluent. But I have been able to collaborate on a textbook that may help carry and communicate some of their language and culture, so that Umóⁿhoⁿ children can continue to learn and live in their people's way. The Umóⁿhoⁿ Elder women speakers I know express this hope for their grandbabies and descendants, and if they want to share the gifts they were given, I want to help them.

I was sitting in my new office, back home in central Kansas, editing that textbook manuscript when I read: "The word for 'old,' itʰóⁿthiⁿadi, breaks down to 'in front,' because the Umóⁿhoⁿ cultural metaphor of the past is in front, where we can see it, not behind us as in English." I'd probably heard this before from my teachers, but now I started to feel the roots of this word actually sink in.

"Itʰóⁿthiⁿadi," I said aloud. How long ago was itʰóⁿthiⁿadi last said here in the riparian zone along the Smoky Hill River, a river whose name in English seems to be a literal translation of the Umó^{n}ho^n name? How might people mill about in this place with the past ahead in view, with memory living and shaping their common sense?[3] How might prairie roots sense paths ahead in the past, in the soil communities they interact with, to make and remake ancestral grounds?

*

All I can do as a writer, it seems, is to sink into the specificities of the few places and plants and peoples I know. Roots help hold me to Earth. From here, I can imagine nurturing a different inheritance for those who come next.

John Weaver, the Nebraskan ecologist who reported on the rates of Bluestem loss during the Dust Bowl, dug trenches and made monoliths to study prairie roots. An herbarium on campus displays the Little Bluestem roots that might have hung over his death bed. Haunting and beautiful, extracted and exquisitely arranged, they feel both harmed and cared for. They feel true.

Maybe what we need is a movement from the clean line of the plough into this tangle of relationship and responsibility. Maybe "creativity" is softening into the complex process of change from which any newness can emerge. Maybe putting the past in front of us will help us make a future in our home globe, in this ecosphere. There is so much beyond my capacity to know. Others will judge what kind of ancestor I will be or already am. But I will continue to celebrate the plants and soils turning sunlight, rain, and rock into food. Like the Bluestem I turn and return to, I'd like to be remembered as grounded.

*

Remember that ground both is and isn't solid, so grounded both is and isn't a stable state. Ground is more than a container, an environment for an organism, or the background or setting for the story's actors. Ground is the soil community, which is not mine to stand upon and possess. Ground overflows boundaries. So grounded is more than obedience to gravity. Grounded is recognizing and rejoicing in the reality I enter just by existing here, tumbling and twisting to interface earth and sky, flowing in feelings of beauty and pain.

Ongoing systems of violence have disrupted many relationships between human communities, nonhumans, and places. People of all kinds seeking positive change are often invited to help renew those older relationships, such as between humans and the plants who feed us, as well as to create new relationships. Philosopher and activist Kyle Whyte (Citizen Potawatomi Nation) brings increased precision to such restorative efforts. In Anishinaabe intellectual traditions, Whyte identifies particular *qualities* of reciprocal responsibility in relationships between various humans and more-than-human beings, qualities that attach social forms and ecological places. Qualities of "consent, diplomacy, trust, and redundancy," for instance, are what "facilitate interdependence in ways that make it possible for the types of relationship to actually have the capacities to achieve social outcomes, including freedom, sustainability, cultural integrity, [and] economic vitality."

Whyte continues, "These qualities of relationships can see societies through some of the toughest of times, which means they support self-determined adaptive capacity that avoids reasonably preventable harms—that is, they support collective continuance."[4] Maybe what kind of ancestor I would like to be is more appropriately considered in the plural potential: What kinds of ancestors might we newcomers need to become here? Seeking to ground our relationships, I give thanks for enduring communities who hold the qualities of whom we might help to move round back, who might help all learn and prepare for the paths ahead.

*

What persists is chaotic and patterned, resting and restless. One spring at Spring Creek Prairie in Nebraska, even the dead Meadow Vole moved.

Around the same time that I was a student starting to feel the absence of Bluestem roots in the basement classroom, I left campus to scout this nearby prairie as an upcoming field trip site for a class I was teaching. The ground plums and wild strawberries were in bloom, purple and white, soft too. I stepped around the flowers. What looked like a rumpled mouse lay back, belly up, tail tense, mouth open to show two teeth. Grasses rustled and a fly zoomed as I approached. I opted to leave the still life be. I was on my way to the badger's hole to check for another claw track. A few weeks prior there was an unforgettably legible mark in the mud there, but today the ground's surface was smooth. Eventually I circled back on the path. Only then did I spot movement at a distance, movement distinct from the wind.

The Meadow Vole's body throbbed a bit, the tufts of fur pulsing up as if a breath struggled through. Matted grasses spread dry under the sun and I came close again, peering but not ready for the shock of dusty black, a fat healing beetle working out and then back under. What remained of the ancestral soil community was re-earthing. The method and motive of the ground is regenerating, including us.

*

I used to think that by writing I could carry knowledge from the past forward in time, like knowledge could be contained in a bucket and passed ahead through creaking chains of words. Like the point of learning was to move knowledge, to give back what had been taken away. Like I was just responsible for my own small shuffling part in the land's longer dance.

That was all good and tough enough. And now there's more. I'm not sure I can carry anything without changing it and without it changing me. Whatever I attempt to just help transmit becomes translated. Take the textbook, those stitched together syllables and lines, cross-hatched with power and meaning, sedimented into a seemingly solid product. I thought I had just been one of the many who helped words like it^{h}ó^{n}thi^{n}adi be carried within the book's beautiful cover. And then there was more. There was a dinner on the reservation with Umó^{n}ho^n Elders and their relatives to celebrate the publication, and ceremonially bless the book. I sat in community—more precisely, I sat in the bingo hall at the casino positioned there, right there, at the table near the head table, in between a colleague and an Elder speaker, looking to my right, my body turning in a clockwise direction from northwest edging slightly eastward so that I could see, as the smoke rose and the prayer was said. What I saw was a translation of the past ahead, the grounded glint of what is usually hidden but always supporting us coming into sight.

Months later a Great Blue Heron joined my husband and son and me on the field during our afternoon soccer game at the park. Stood close enough, just long enough, to make clear how a heron is still taller than my growing little boy. (Little Bluestem is still taller if you count what grows below, which I do.) What my child, what the Heron's grandbabies and descendants, will weather felt like a heaviness in my chest, a force that drew my breath in so sharply that I thought of my eventual last breath, my probable burial. The Heron, meanwhile, took flight.

*

The seeds I have picked up to share and repeat are words of encouragement, shaped to incite courage. May we. Let us. We shall. Care for.

The tangle has stayed with me, the Bluestem whispering as I listen, winding lines of curious inquiry, wreathing downward.

A gentle unsettling takes root as common sense.

ESSAY

My Home / It's Called the Darkest Wild

Sean Prentiss

I. SOLSTICE LAKE

January, Solstice Lake is frozen. Snow sleeps upon looming Solstice Mountain. Clouds float pregnant across this Vermont sky, offering no beginning, no ending.

Into this world a daughter will be born. In two weeks, Sarah Eve and I will carry a cooing girl down a snowy hillside toward a home where chimney smoke corkscrews. This baby, I imagine, will love winter because this baby will have sprung from Solstice Lake's whipping wind, beside Solstice Mountain.

II. LENAPEWIHITTUK RIVER

Thanksgiving, and my mom, Grandy; my nine-month-old daughter, Winter Eve; Sarah Eve; and I wander beside Lenapewihittuk River. After the bustle of Thanksgiving, we talk and talk before being quieted by the roil of the Lenapewihittuk.

The trail we walk snakes downriver from where I grew up. In my homeland, directions are upriver, downriver, or inland. How else can a person tell directions other than by the flow of a river? This region is river valleys and some of the world's oldest mountains—the Kitahtëne—grandmother and grandfather peaks that hem us in.

III. SOLSTICE LAKE

Middle January, I cradle newborn Winter Eve against my chest, skin to skin. Outside the home with chimney smoke coiling, winter coils itself around trees and wisps across Solstice Lake. Plumes of snow, plumes of smoke—a funnel from home to heavens.

I read this baby poems to soothe her. This dawn, I choose Cold Mountain[1]— that ancient Chinese poet who lived in a cave on Cold Mountain and wrote his *poems on rocks* (180) and *on cliffs* (268)—because the poet Cold Mountain might explain to this baby (and to me) how to live beside Solstice Lake, beneath Solstice Mountain, in a home with chimney smoke spiraling. Cold Mountain's poems might illuminate to this baby (and me) how to turn place into home, how to turn home into self, how to knot self to lake and mountain like the poet Cold Mountain sewed himself into the mountain named Cold Mountain.

When Winter Eve cries, I read Cold Mountain aloud: *What's this crying for / those tears as big as pearls* (74). When this baby cries for Sarah Eve—*Her beauty transcends the immortals* (20)—I read Cold Mountain aloud.

IV. LENAPEWIHITTUK RIVER

Sarah Eve, Winter Eve, Grandy, and I walk along this November river, which, if Cold Mountain were hiking with us, he might say, *The floodplain river is wide* (223). Sarah Eve cradles Winter Eve to her chest, enveloping her. Winter Eve should be napping. Instead, she gazes up at sycamores rising so high they feel as if they circle back down upon us. Hemmed and held by trees. If Cold Mountain was here, maybe he'd write, *Yellow leaves fall / white clouds sweep / rocks are huge / woods are deep* (253).

The four of us scatter ourselves downriver, a direction others might call south, away from my childhood home. We watch as the Lenapewihittuk, growing more violent, tumbling toward ocean, batters rocks, writing a poem of its journey upon earth, carving stanzas into soil, *grinding away the rock of the earth* (227).

Winter Eve silent but awake, which is how she is in the woods. The woods shush her pearl tears. *This is such a restful place* (225), I hear Cold Mountain whisper. On days like today, Winter Eve needs nothing but river-song and tree-sway.

*

V. SOLSTICE LAKE

I was not born on Solstice Lake, nor was I born in Vermont.

I was born beside mountains older even than Cold Mountain, the mountain that the poet adopted his name from. And I grew up beside the Lenapewihittuk River, which is older than the last glacial period.

It took me twenty years (half a lifetime so far) to find this Vermont home, but once I knew it existed, I raced to Solstice Lake as quickly as I could. So did beautiful Sarah Eve. Now we plan to never leave. Cold Mountain, he got to Cold Mountain when he was thirty. Sarah Eve and I were slower. But, now we vow to make Solstice Lake our heart-home. Solstice Mountain will be our shoulders: *Towering cliffs were the home I chose / bird trails beyond human tracks* (1). If you are *looking for a refuge / Cold Mountain will keep you safe* (2), Cold Mountain wrote. Solstice Mountain will keep us safe, Sarah Eve and I whisper.

VI. LENAPEWIHITTUK RIVER

I was born beside this ancient river. I spent twenty years (half a lifetime so far) beside Lenapewihittuk River, before abandoning it. I abandoned it not to run from family or the Kitahtënes or home or river (all of those terms, *family*, *the Kitahtënes*, *home*, and *river* are words for the exact same thing, for the community where one's soul resides) but to run from highways that sprouted like wild chervil, subdivisions that rose like wild parsnip. All of it (the highways, the subdivisions, the invasive weeds) angry and vengeful and without an ancestral plan.

So for half a lifetime, I ran through sixteen states and few but this one became anything more than a fleeting home.

And Winter Eve, she was not born beside the Lenapewihittuk River. She was born beside a glacial lake, a kettle pond, six hours north of here, Solstice Lake.

> Still, she is of water as I am of water.
> We are merely from different waters.
> When I hold her in my arms, our waters confluence.

VII. SOLSTICE LAKE

Beside Solstice Lake, Sarah Eve and I sculpt land not into the shape of ourselves, for we are here for little more than a moment, but into the shape of Solstice Lake, so that our land mirrors back itself in our summer lake—a spectacular re-imaging of our home upon water.

It is as Cold Mountain writes; we should attire ourselves from the land *with bark hat and wooden clogs* and *with hemp robe and pigweed staff* (207). So beautiful Sarah Eve, Winter Eve, and I wear gloves of garden compost, lips painted red from raspberries, tongues sticky from honey, eye lashes glistening with cold smoke snow.

VIII. LENAPEWIHITTUK RIVER

Beside Lenapewihittuk River, Grandy turns to me and says, *This is your ancestors' home.* I know this story well: German river people traveling in the 1740s from the Enz River across an ocean to the Lenapewihittuk, to these Kitahtënes Mountains, which back then homed an entire nation that is now gone or dispersed or lost because of my ancestors.

Grandy pauses beside a cellar-hole from the 1740s. Much of this history has already been swallowed. Only these few rocks remain. We look at these river rocks, rounded by the wearying of Lenapewihittuk's current, piled into a square. From that generation to today, eleven generations from Johannes Georg Beck to Winter Eve Prentiss. The names are a genealogical poem we sing:

> From all who lived here first to
> Beck and Gucker to
> Beck and Weidman to
> Emery and Beck to
> Miller and Emery to
> Miller and Everhart to
> Searles and Miller to
> Hall and Searles to
> Hall and Peters to
> Prentiss and Hall to
> Prentiss and Hingston to
> Winter Eve Prentiss, just
> a cooing baby girl, all but
> the last born beside
> Lenapewihittuk River.

This homestead is empty and has been for maybe two centuries. Moss-covered rocks and sycamores growing from the center of what used to be the home. The roof has become soil that has fed some tree that has long since fallen and turned itself back into soil that feeds this tree that grows from what used to be the roof of a home.

The poem of our home beside Lenapewihittuk River is written in sycamores that outlive us, that root themselves back to our ancestors.

IX. SOLSTICE LAKE

No matter how permanently Sarah Eve and I build the home with chimney smoke corkscrewing, Cold Mountain prophesizes, in some distant future, that *its sides have caved in / its walls are cracked / its beams are askew / its tiles lie shattered / its decay won't stop* (181).

By then, Winter Eve, beautiful Sarah Eve, and I will be bones swallowed by earth. Our lives *blooming at dawn gone by dusk* (262). Cold Mountain tells us of death: *Once they are buried beneath the weeds / the morning sun is dim / their flesh and bones disappear* (52). Cold Mountain tells us that *births and deaths never cease* (112). Cold Mountain tells us that *Death remains impartial* (50).

I lullaby Cold Mountain to this crying child: *both life and death are fine* (100). I cry as I read to her.

X. LENAPEWIHITTUK RIVER

Sarah Eve, Winter Eve, Grandy, and I stand beside this ancestral home that is now piled river rocks. What were the walls? Logs now rotted? Was this smaller square rock hole the outhouse? All of it caved in but the river rocks.

If we were to hike eight miles upriver, Grandy could sit us before a small tombstone. We would find a stone protruding from the ground. On it is chiseled an ancestral poem.

Johannes Georg Beck
May 1705—May 1775

Sarah Eve holds Winter Eve in her arms eight miles downriver from this cemetery. I feel my own bones (middle-aged and wearying). I stand beside Grandy beside our ancestral cellar hole and see life's poems written on Grandy's face. Her hair wild poems that speak in this autumn breeze.

XI. SOLSTICE LAKE

Cold Mountain (maybe the mountain and the poet this time) shows this child and me that *the seasonal round never stops / one year ends and another begins /*

ten thousand things come and go (23). Cold Mountain shows us that we are all refugees until we are home, until we become home, until we defend home as we defend family (for Cold Mountain is a mountain, Cold Mountain is a man). Those are one and the same.

Let us write the poems of our lives on this land. Let us let the lands and waters of our lives write their poems upon us.

XII. LENAPEWIHITTUK RIVER

Our ancestors, the ones who stacked rocks into stone rows, who heaped rocks to create cellar hole foundations, if they knew anything, and surely they did, they knew that no matter how far they moved these rocks, soon enough the world would swallow the rocks back into its throat.

These mountains, the Kitahtënes, once rose to the tallest in the world. They now are just hunched hills like a grandmother aging shorter. Still, these mountains have sheltered my ancestral family since the 1740s and are as much our grandmother as Grandy's mother, Mommy Jo, was. The elder watching, wisely.

Ancestors rise and ancestors fall. Now they rest within this soil (becoming this very soil) beside the Lenapewihittuk and soon enough too we will sleep within this soil (or some other soil) for a person can be nothing more than the soil and water they came from, and we all come from some exact soil and water, which we might call earth-home, water-home.

XIII. SOLSTICE LAKE

Or else it's the lesson that Cold Mountain never wrote down that teaches us how to live this one life, how to move from the realm of decedent to the realm of ancestor.

Cold Mountain (the poet), when it was time, simply slid into a crevice on Cold Mountain because his *true home is Cold Mountain* (177). Cold Mountain became Cold Mountain. Maybe Cold Mountain sang this song as he entered his final home: *My home / it's called the darkest wild* (55).

XIV. LENAPEWIHITTUK RIVER

Sarah Eve, Winter Eve, Grandy, and I stand before that original Pennsylvania homestead. I think about ancestors. Cold Mountain writes, *All the people I see / live awhile and then die* (223). Now there is so little left here but (tomb)stones jutting from earth, whispering earth-stories. A few piles of river rock shaped into what we must imagine into home, barn, outhouse. There is so little left but stone rows edging what used to be wild woods turned into small plots of crops returned now to wild woods lined by stone walls that serve no use other than to tell an ancestral story, which maybe is that nothing from our hands lasts longer than just a fleeting breath.

*

XV. SOLSTICE LAKE

Solstice Lake is ten thousand years old. Solstice Mountain is weathered from carve of ice, battering of rain, cleave of snow.

We, the three who live in the home with chimney smoke twisting, arrived so recently that we are as unnoticed as a water strider on the lake. When we die after tomorrow's tomorrow, all I ask is *Let heaven and earth be my coffins* (15).

Once we sleep within coffins (*none escaped the Wheel of Birth and Death* [237]), Solstice Lake shall still fill up with the world's rain and snowmelt. Solstice Lake will overflow into Solstice Creek that churns its way toward a river that churns its way to a big lake that churns its way to a river that churns its way to an ocean that is filled, too, by the Lenapewihittuk.

Solstice Lake cares for us as much as it cares for a raindrop landing upon its sheen.

But maybe that is all wrong. Maybe the raindrop is the ancestor, the lake is the living, and we—Sarah Eve, Winter Eve, and I are the descendants, and the rain teaches us that it will fall, run with a river, and then return to the heavens, a cycle forever.

XVI. LENAPEWIHITTUK RIVER

Today, Lenapewihittuk River, born before the last glaciers, continues to carve this Pennsylvania world in the same way the Beck family gnawed a cellar into Pennsylvania dirt.

We (the four of us—spanning three generations) stand beside a cellar hole eleven generations before Winter Eve. Now those ancestors are gone, buried in earth's coffin. Next to join them, my mother and then me and then Sarah Eve and last, precious Winter Eve. When we go, the river will still gnaw its way onward.

If we learn anything from river rock, it is that the river will remember us as much as it remembers a pebble it caressed moons ago. And in that new home on Solstice Lake, once we are gone, first me, then Sarah Eve and last, precious Winter Eve, our house will crumble to even less than river rock cellar holes. Just twelve cement tubes rising from the earth and the slow drip-drip-drip of whatever poisons make up T-111, insulation, and drywall.

XVII. BESIDE EITHER WATER

Today, wherever I stand, *I sing this one song* (339):

> We are nothing but descendants of river and lakes.
> We were birthed from soil.
> Soon enough, we return to soil.

XVIII. BESIDE EITHER WATER

But what do I know?

I have only just begun this journey into being an ancestor. This short life, I have learned so little. Barely yet a father. Only a man who ran from one water to the next.

So don't listen to a word I write upon paper or rock until I have slipped through a crevice on Solstice Mountain and disappeared forever into this mountain. Not a moment before then.

This long song I just sang to you, as Cold Mountain writes, is nothing more than *a blind man's song about the sun* (241).

INTERVIEW

Wendell Berry

An Inheritance: The Survival of Human Love in the World

Leah Bayens

BAYENS: *Conversations about ancestors invite a broad understanding of those who came before, but I'd like to start with some ancestors very close to you: your parents. Will you tell me about them and what you learned from your parents about being responsible to a place?*

BERRY: My father's dedication to this place was elaborated by him in principles very close to him personally because of our family's history. His devotion of so much of his life and effort to the Burley Tobacco Growers Co-operative Association can be taken, I think, as the sign of his commitment to his own place, which then by imagination and sympathy he applied to other places in the so-called Burley Belt. He knew what they were going through and what they were up against. I think that made him a public servant. My mother was a quieter person, but she was a trained pianist and a devoted reader of books. She was an accomplished flower gardener, and she loved the wild flowers and birds. The first field guides that I ever saw were hers.

It would be hard to say which one of them I am most indebted to, and I wouldn't even try to say. My dad was not a reader except for the things that immediately mattered to him. He was a very good reader, a very close reader, a useful reader for anyone who needed to be read closely, as I did. He could help. My mother was a more literary reader, and the story was that she read the poets to me from infancy on, I reckon, until my brother was born.

BAYENS: *Were your mother's field guides ones she picked herself, or were they passed down to her?*

BERRY: She had gotten them herself. I still have them. To me, that signified a permission to take an interest in the birds and flowers, just as her reading signified a permission to read. My love for the outdoors was much stronger than hers, I think, though she was a flower gardener all her life. This issue of permission is very important for a young person. You get the permission, you go with it. After that your motive is your own.

BAYENS: *Do you think the permission you got from your mother to explore outdoors, to explore the place where you lived, was different from the permission you got from your father?*

BERRY: Oh, yes. I think so. My father was a taskmaster in a way that my mother wasn't. My mother, poor thing, for fear that I would drown could discipline pretty harshly with a switch, and I know that it hurt her worse than it hurt me because for me the pain didn't last very long. My father's metaphor was always the work team. He didn't want me to have "slack in my traces" or to "lie down in the harness." He'd go back to that many times. He had experienced very closely the economic struggles of his parents. His parents experienced many years of hardships because of the economy. The 1890s was a depression. The aughts at the beginning of the twentieth century were the decade of the disastrously low prices paid by the American Tobacco Company and James B. Duke. The 1920s was a farm depression. Country people went into depression ten years before the Great Depression came. My dad knew all that and felt it all. That established his fear for his children and in particular me; to have a writer for a son was pretty much a challenge, as I now know that having a writer for a son would have challenged me. I would have been as scared as he was. He was a taskmaster.

He was also, along with that, an excellent teacher. I'd be going away to school, and he'd take me with him to look at the cattle, mostly steers in those days. He'd say, "Now pick one, honey. Pick one, and look at it closely and see its condition. When you come back, look at it again." Well, I never did do that consciously, but you see the point. That is a basis of sound criticism. It's Poundian criticism: put the slides together and compare.

My mother had a good mind. She was an intelligent woman, well-educated. My daddy's mind was excellent also, and more active. When his mind went to work, it was work. He had done a lot of comparing good work to bad.

BAYENS: *Knowing how to tell the difference is—*

BERRY:—the work of criticism that we're all called to. When I published *The Unsettling of America*, it was a revelation to him. It showed him a competence that I had, but it also showed him that I was his ally. He was sick, and I went up and sat by him. He was in bed—reading my book. He put the book down, and he said, "Well, if you never do anything else, you've done something."

BAYENS: *That's high praise.*

BERRY: Yes, it was. Let's see. That was 1977. I was 43.

BAYENS: *Talking about ancestors requires memories, some knowledge of history, and imagination. I'm curious about the ways we hold on to history and also how it's lost and how it's skewed. I keep coming back to your poem "What Passes, What Remains." In that poem, you talk about how the marks of history are made in the landscape. The poem begins, "Here the mingling of the waters / of Shade Branch, Sand Ripple, / the dishonored Kentucky River, / tells the history of our country / which is the history of our people." What have you learned about our ancestors from the marks on the land? And the corresponding question is: what kind of marks do you think you've made?*

BERRY: The marked land is a significant study, and we're not very good at it. Dr. [Thomas D.] Clark, in one of his late interviews, talked about our great need for a history of the land. We don't have a good history of farming in Kentucky. We don't have a good history of livestock breeding and production. A history of the land would complete our understanding of the history of the people. It would be indispensable to understanding the history of the people.

I can talk a little about my father's father. When he got ahold of the Home Place—bought his siblings' interest out and owned the Home Place, what we used to call the pond field was gullied out. The pond is still there. My grandfather dug it with a team of mules and a plow and a slip scraper. The plow loosened the dirt so that the scraper could pick it up. A slip scraper was like a big shovel with two handles, guided by a man and drawn by the mules. My grandfather took the earth that he dug out of the pond and used it to fill the gullies in that field. But under economic oppression that was so constant in his life, he planted a field in corn—more than he should have. A big rain came and washed the field. He spent the rest of his life making amends for that.

I think that stayed with my daddy. I think it made him and then me a conservationist. I remember driving with him and looking across at the farm adjoining the Home Place. The ground had been broken way too low on that hillside. When the farmer had gone out of the field with his harrow, he went straight up the hill, leaving a mark that would be deepened by the runoff. I remember my father's indignation at seeing that. His metaphor for a gall, a sore, on the land was a harness gall on a work animal. When it was healed, he would say, "It's haired over." The metaphor acknowledged the livingness of the land. "You've got it haired over." The land is an organism. It's a thing that can be hurt, and it requires sympathy. It requires an elaborate comprehension, too, and a commitment. There's passion in that, an identification.

For people whose minds are formed from an agrarian or agricultural point of view in one place, that's a distinguishing thing. The land, then, is the basis of the mind. Your own story there, the story you know, of human life on that place gives the mind some of its most significant contents. I got that very much from my dad.

BAYENS: *[Your son] Den told me that the hillside next to your house hasn't been mowed in nearly three decades.*

BERRY: That hillside is a significant agricultural exhibit. I wrote about that for *Farming Magazine* in a story of our sheep flock. About 1990, I was pretty well stretched out by work and commitments of various kinds. By then, we had been on this place 26 years, and that hillside had been mowed every year. For most of those years, I had mowed it with a team of horses and a McCormick High Gear No. 9 mowing machine. A good thing to do. Very pretty work on a hillside.

BAYENS: *It's very steep!*

BERRY: Yes, well, Den says it'll make you grow hands on your behind.

BAYENS: *He would say that.*

BERRY: About 1990, I said, "I'll quit mowing it and see what happens." I had a kind of vision. If I withdrew the blade, then the woods would come back. But I kept putting the ewe flock up there in the wintertime, and the woods never came back. What I learned is that if you put the sheep on the hillside in the wintertime, and they become hungry for something succulent, they'll bite off even the cedar sprouts. That says to me that if you have a diversified farm, and you've got sheep with your cattle, a lot of investment in pasture clipping can be dispensed with. There that hillside is now, and it's going on thirty years. Last summer, we killed a couple of big old cedars that had gotten started up there, but that's all. All these lots around our house have not been mowed for at least that long, maybe longer. We're using our meat animals for the work of land maintenance.

BAYENS: *I recently reread* The Hidden Wound, *in which you write about trying to understand your family's and our neighbors' stories—what they revealed about race, power, affection, and membership. In one passage, you wrote about storytelling as a tie that binds generations. There's a kind of ritualization to storytelling, and yet there were some stories that were "casual recollections" about slavery. You wrote very earnestly and, to my mind, very vulnerably:*

> There was a peculiar tension in the casualness of this hereditary knowledge of hereditary evil; once it begins to be released, once you begin to awaken to the realities of what you know, you are subject to staggering recognitions of complicity in history and in the events of your own life. The truth keeps leaping on you from behind. . . . (5–6)

When stories are told uncritically of this kind of "hereditary knowledge of hereditary evil," one tack is to cast aside the stories and even the storytellers as irredeemable. But another tack is to sort through the history and to rout out the root causes and contexts of entwined cultural and ecological problems, sometimes devastations. I think that's the approach you've taken. How have you come to terms with your—our—predecessors' shortcomings and offenses, and how do you figure out what to salvage and what needs to be chucked?

BERRY: I don't think you're justified in willfully chucking any of your past. This monument business is a very good example. My brother John once told me, "You can't improve your history by hiding the evidence." If it's there, you have to deal with it. But this is a very complex thing. My sense of the complexity is much increased. I wrote *The Hidden Wound* fifty years ago, when I was 34. There was a whole lot I didn't know, and I was a little too obligingly feeling guilty, I think. Maybe there *is* such a thing as hereditary guilt; no doubt there's such a thing as original sin. We were born here, and there is something wrong here, however you want to phrase it. Putting it to myself now, once I went into the sin business on my own, I didn't need to inherit it. I kept myself well enough supplied with things to feel guilty about. Guilt as a motive is as bad as anger. The solutions have just got to come from somewhere else. My understanding is that it has got to come from love. Everything else is going to lead us wrong.

This business of the past involves the historical record, the formal histories written by trained historians. There's the documentary history, and we are obliged, whether we are historians or not, to paw through those documents if we're going to take the past on as an obligation. Then there's memory. Inevitably all through it is imagination. To pick those strands apart is a big job. At some point you're going to need fiction, which is simply to put things in order. It's a relief to think of it in that way. Some of the connections are going to have to be imagined if they are going to exist at all. There's this ongoing responsibility.

The historian John Lukacs has been very helpful to me in this way. We should understand our responsibility not as issuing from a substantially known or reliably known or proven history; it issues from the understanding that the history we've got is provisional. It's always going to be provisional. Lukacs says that there is no knowledge but human knowledge, which is a properly humbling statement. There's knowledge for us, before we know it. Then we're responsible for it and for calling it "truth." He says that knowledge is neither objective nor subjective, but "participatory." We participate in history by learning it and by living in it. So what I do with what I know of the past is a test of it and in some sense probably a revision of it. I heard Lukacs talking to the students at the McConnell Center at the University of Louisville. He was making them very nervous by saying that even the science that you learn is subject to change. They asked, "What is the curriculum?" "The curriculum is," he said: "Is that what you *really* think?"

So you see, this is a relationship. It isn't something to be learned by rote. This is the issue about ancestry. It's not a dead relationship. History is not an autopsy report. The difference, which my friend Kathleen Raine upheld in her work, is that you go to your literary ancestry to learn *from* them, not *about* them. You don't go to Shakespeare to learn *about* Shakespeare. You go to learn what he has to teach. Chaucer. Dante. Homer. On back. Young people need, above all, to claim that ancestry. If we want a tradition, then there's no place that we Westerners can go except to our Medieval teachers. We're still in disintegration from what you might call the explosion of the Renaissance. We didn't just progress. We progressed at a cost. C. S. Lewis says that for every nail we drive in we drive another one out.

BAYENS: *As you know, we're starting an agrarian, liberal arts education program here in Henry County. What do you think a proper education ought to pass down to students?*

BERRY: I think we have to start thinking about education with the assumption that it's provisional. Education is provisional. There's lots of looseness in it. The education specialist will try to control that by resorting to a method. But when you get to the actual students, it's another order of business. You find yourself dealing with grammatical and syntactical problems for which there are no names. And students who are smart or stupid in new ways, and so on. Here are people with a need that the curriculum has assumed, and they're there from no consciousness of their need. The stakes are very high. The predicament of the teacher is largely the same as of a parent. In spite of all the history that we can learn or know, time is always starting from right now. In some ways, history may seem to repeat itself, but there's always something that assures we'll be unprepared. Your own lack of preparation to teach is going to be one constant you can get at. The often unprecedented neediness of the young is another constant. You're reduced to doing the best you can with what you've got, which is awkward.

I think I transferred what I learned in the tobacco patch to what I learned in college.

BAYENS: *How so?*

BERRY: Keep at it. When you've got the task to do, do it the best you can with the ability you have. There isn't any way around that. You have to go through the work, not around it. Over and over again you discover that when there's a crop that has got to be put in.

BAYENS: *There's a beginning, a middle, and an end. And then you start all over again.*

BERRY: That's right. Then the work becomes governable by what you've already learned about it and by your passion for it, your liking for it, your affection or love for the people you're doing it with. And then there's always more—a generosity implicit in our world. There's always another year to try to do better in, more to learn than you've learned so far, more to read than you've read so far. But there's something else, too, and you have to become eligible for this in some way that I don't understand. I've checked this with other people who confirm it. Why is it that you've found the books you've needed just when you've needed them most? Over and again. How is it that you found the friends you needed when you needed them most? It's happened to me.

BAYENS: *The concept of membership is something you've given a lot of thought* to, and as a result I've thought about membership—but maybe it's not fair to put that on you.

BERRY: It's fair enough if we understand the context of that statement is provisional. It happened. I came upon it because I knew the passage in St. Paul: "We are members one of another." I had that in mind. Then Burley Coulter goes a step beyond St. Paul out of my knowledge of ecology. "We're members of each other," he said, "all of us, everything. The difference ain't in who is and who's not but who knows it and who don't." That's a part of this happening that I'm calling provisional. It's not finished. If we thought that you wouldn't have come on that except by *me*, that would unfit us for continuing the conversation. You would have gotten it from somewhere. I would have gotten it from *some*where, if I wanted it. Somewhere there is this word *membership* in our language. There is such a thing as needing it. The need for it puts you into a task. I don't know what all is involved in this, but I don't need to accept an embarrassment in that statement, in short, because I know that it's all been a gift to me. I didn't originate any of that stuff I have written. One way or another, it's been a gift.

BAYENS: *On first glance, to talk about ancestors, is to talk about the past, and of course it is. But it's also to talk about the present.*

BERRY: Because they're here.

BAYENS: *That's right. It carries right along. For some people, it's to talk about that thing that doesn't exist yet and might not come to be, which is of course "the future." Many people spend perhaps an inordinate amount of time thinking about the future. How does that impact who we become and the legacy we leave?*

BERRY: Well, it's a terrible kind of isolation, for one thing, because you're then alone with your fantasies—whether they're fantasies of inevitable progress and improvement or fantasies of inevitable collapse and destruction. Nobody's fantasies are like anybody else's. It's a very effective way of cutting yourself off from everybody else.

BAYENS: *Taking all of this into account, what kind of ancestor do you hope you will be?*

BERRY: I wish I could be an ancestor who would give permission to love things and to mean it and to understand what meaning it means, in all its complexity and difficulty. I would like to do that. I would like to give my successors permission to be critical—to ask the questions. Mostly, it has to do with the survival of human love in the world. This is not a simple thing.

Allen Tate says there will always be some who remember. He says the main work of the civilized mind is always going to be salvage. You'd hope that by teaching literature to your students, you wouldn't teach it as a cut-and-dried thing to be known about. They'd miss the passion in it.

So you say to your children or your students, this is something, my children, that you may well need one day. You're telling them, in a way, that there's such a thing as a long fuse. The light, or your need for it, may come long after you've been here with me.

BAYENS: *But you'll at least have some grounding.*

BERRY: Down the road a ways, they'll say, "Oh, that's it! That's what she was talking about." You learn "Take no thought for the morrow," when you're a kid. For some reason you remember it, maybe, but it doesn't register much. Sometime, then, it finally comes to you that thought for tomorrow is a great waste because you can have *any* thought for the morrow. First thing you know, you'll be wishing, and who can be trusted with that?

We had a gathering at David Kline's place one time, and I overheard a bunch of young Amish men who were all in a group laughing. What they were laughing about was that their fathers had turned out to be right.

BAYENS: *And that was hilarious to them.*

BERRY: Yes, it was a great joke. Mark Twain told a story about his father. He said, "When I was a boy of fourteen, my father was so ignorant I could hardly stand to have the old man around. But when I got to be twenty-one, I was astonished at how much the old man had learned in seven years."

BAYENS: *Thank goodness he'd had that son to teach him.*

BERRY: [My neighbor] Owen Flood once told me, "They'll never be worth a damn as long as they've got two choices." I've lived into that, but I've never gotten to the end of it. Several things he told me I haven't gotten to the end of. But finally I realized that one of the things that culture does is to remove the second choice. That's what we call commitment.

BAYENS: *There are certain approaches to education that encourage a kind of the-world-is-your-oyster way of thinking. You can be anything.*

BERRY: It's a terrible lie.

BAYENS: *It's a terrible lie, and it's too much pressure. Who on earth can manage all those choices?*

BERRY: It's driving the young people crazy. I think it's literally driving them nuts.

[Tanya Berry joins the conversation.]

TANYA BERRY: That may be part of what the problem is these days with addiction and so forth. It's just total confusion for a kid.

BERRY: I began to think of this years ago. The Rodale people, when I had an interlude with them, had a farm out by Kutztown. They had an old order Mennonite farmer next door, Ben Brubaker, and my friend Jim Foote had been working up there. He paid a lot of attention to Ben Brubaker. He said, "You know, Ben's sons were growing up, and I began to notice that he was disappearing. He and his wife were going off on little vacations—a little longer every time." Ben's sons weren't twenty yet, but they were full grown people. They knew what to do.

BAYENS: *They knew how to be responsible.*

TANYA BERRY: You all were talking about the future. I was thinking about my least favorite hymn in the hymnal: "Turn your eyes on Jesus. Look full in his wonderful face and the things of this earth will grow strangely dim." But that's what's wrong in religion. If that's all you're thinking about, everything grows dim *here*. Getting ahead in your work, or whatever it is keeps you from seeing.

BAYENS: *Yes, from seeing the work that's in front of you or the lives that came before you.*

BERRY: There used to be a set of stories that I think were in McGuffey, but I have never looked them up, about the Three Sillies. The one I always remember was about the young daughter who's sobbing uncontrollably. Her mother had just sent her to the cellar. She looked up and saw a heavy hammer hanging over the cellar stairs. She was weeping, weeping, weeping because when she got married, she might have a son, who would grow up a fine young man. And her fine young son would be sent to the cellar as she had just been, and that hammer would fall on his head and kill him. Much of the future could be taken care of simply by unhanging the hammer—the right thing in the present.

Wes Jackson made up his mind when they'd sited a power plant for Western Kansas, because nobody cared about Western Kansas and hardly anybody lived there. They were going to build this huge coal-fired power plant out there, where there wasn't any coal, of course. Wes decided against it one morning, and by night he had this movement started to oppose it, and they did oppose it. The most telling thing about that story is that they never mentioned global warming because that would have divided their constituency. They talked about waste and pollution, which the people opposed. They knew what the right thing was. They beat the power plant. My emphasis has shifted. I'm not going to be baited anymore by people who ask me what to hope for or how do I have hope. I'm just going to say, "I know the right thing to do." That gives me a lot of consolation and all the hope I need.

BAYENS: *In certain circles, if you refuse to focus on the future or on big solutions, people balk. I'm convinced that we're getting sidetracked; it's another smokescreen.*

BERRY: Absolutely. I think Mary's showing what the test is. If you're not willing to do the small thing that is obviously right, and is right under your feet, you're not going to be much help.

[Mary Berry is the founder and executive director of The Berry Center, a nonprofit organization in Henry County, Kentucky, that advocates for farmers, land-conserving communities, and healthy regional economies. The Wendell Berry Farming Program is one of The Berry Center's four primary projects.]

SEPTEMBER 26, 2018
*Lanes Landing Farm
Port Royal, Kentucky*

POEM

To the Children of the 21st Century
Frances H. Kakugawa

How do you keep your fingers so free of dirt?
How do you come in from play without
Mud on your feet, your clothes, your cheeks?
How do you not even sweat?

How do you speak without giving eye contact
To the person sitting in front of you?
How do you spend time with your friend
Without conversation?

Oh Children of the 21st Century,
Why is there silence in a room filled
With family on this holiday?
How did you become so mute?

Do you know how rain feels
Soaking your clothes to your skin?
The smell of salt in your hair
After a dip in the sea?

Have you watched a little seed
Pushing its first breath
Out of soil you've patted down
A few weeks ago?

Can you see a cardinal, a mynah,
A crow, with your eyes closed, listening

To their signature songs they sing out to you
In your own back yard?

Do you know the feel of your grandpa's grip
Warm and strong in your hand?
The story behind that long scar that runs
The length of his arm?

Do you carry memories
Of your grandma's smiles
Each time you had said,
Hi Grandma. Can I help you?

Do you ever count clouds, lying
On soft green grass, laughing
Over silly stuff shared with a friend?
Do you ever cry over a child starving

In Africa or in your neighborhood?
Stop trees from being cut
For freeways and shopping malls,
Fancy sports arenas?

Have you ever used the eraser
At the end of a pencil,
Writing a poem, a song, a story.
A thank you note?

Do you know the feel of crisp
New pages of a book, as they unfold
Moving plots, faster than your impatient
Fingers can follow your eyes?

Oh, Children of the 21st Century,
How did you become so dead?

* 2 *

Reckoning

Even if it destroys their sense of comfortability, I will be the ancestor who encouraged people to get into the deep, tragic, sometimes incredible melancholy of all that was and came to be.

ERYN WISE

POEM

Forgiveness?

Shannon Gibney

Father, I never met you. Yet you are the cleft of my chin, the clean space of my forehead rising in inquiry, my incessantly chattering brain. Ancestor, he who came before me, he who made me, he who made me possible. I want to say I am part of your line, but I have no idea what that would mean, how my segmented piece might join yours. *Street life got 'em,* was what my aunt, your sister, told me when showing me the parade of photos of your siblings who died much too young. About you she said, *He had emotional problems.* I knew you went AWOL in the Air Force, and that my birth mother said you were using drugs during the short time you were together. *Died from infection that resulted from injuries suffered during a high-speed police chase,* was the last part of your story I tracked down on your death certificate. My best friend and I in a rural Michigan county, lying down in the cornfield beside the county documents office, me holding the remains of your aspirations and afflictions in my 19-year-old hands. Repeating over and over again, *I don't know what to do with this. I don't know what to do with this.*

And what to do with you, Mother? You, too, are ancestor now. Cancer you refused to have treated finally ate away the last of your small intestine. Even in death, I imagine your stubbornness to be a large hill you sit perched on day and night. Unlike him, I found you, met you. Although we were never actually able to *see* each other fully. Or even partially, I think. It still has to mean something though, daughter and mother finding each other again, 19 years after they removed you from me, and you would not allow yourself the relief of crying, or even watching as they took me. I imagine. I know because of the dull pressure in my gut every time I think of me, that baby in the state home alone for five months, with little to no contact or touch. You, every January 30, telling your family, *I'm going to go get her. I'm going to find her.* And them holding you back, convincing you it wouldn't work, that I was a child with my own loving family now, that you had made the right choice in giving me up, that you would find me finally, as an adult. And you did. And I did. And we broke apart and came together, and you rejected me again, and I fell apart again, and you were unkind, Mother, even though I know you were doing the best you could.

If I sit solidly in the middle of this stream of time, I can see myself as I see you, Ancestors. Telling these stories, and then letting them go. Urging your grandchildren to touch the ground sometimes, not just walk upon it. Calling the dust of me to continue to disappear, and yours, too. Going back to wherever it was we came from, wherever we have landed, wherever we will end up. This is what forgiveness looks like to me: Falling back into that stream, together.

ESSAY

Sister's Stories

Eryn Wise

I was 11 years old when Miss Pixley told our Sunday school class that when god forgave your sins, he cast them as far as the east is to the west.

I couldn't wrap my head around her statement so I asked, "How far is that?" She said, "Imagine that god took a single grain of sand off the beach, just a grain, and took it across the world to a faraway shore and dropped it off there before coming back, grabbing another piece of sand, and carrying it all the way over to the same place he'd been before. He'd lay it next to that other grain, and keep doing that until he moved every single grain of sand from one shore to the other; all the way to the other side of the world. . . . *That's* how far away he cast your sins."

I remember thinking, "Why would I believe in a god that spent all his time moving sand to try and prove that he'd forgiven someone?"

I've determined that I want to be the kind of ancestor who was known for her ability to offer reprieve to those that need it, in whatever way my support can be received. If someone had told me that forgiveness could flow like a river from one open heart to another, without the process making the act that needed forgiving so arduous, it would have mitigated much apprehension in my childhood. I come across so many youth in my work who, like myself as a young person, are told that their feelings of anger are invalid, and that they should simply pull themselves up by their bootstraps and carry on. When I'm an ancestor, they'll remember my dignified rage.

I am of the firm belief that adults underestimate young people's ability to comprehend the world around them. We human helicopters hover around our children, often sheltering them from the facets of our societal realities, much to the chagrin of those made invisible by our shielding. Even if it destroys their sense of comfortability, I will be the ancestor who encouraged people to get into the deep, tragic, sometimes incredible melancholy of all that was and came to be.

Sometimes I imagine Viking ships, or Columbus's crew careening onto the shores of Turtle Island. I picture a people that moved upon the land with reverence, whose roots grew from their feet straight into the earth, feeling the shores shake when the travelers arrived unannounced. Our people must've been so frightened at the foreign sights and sounds, and the more I think about what happened to them after those boats dropped anchor, I feel sick inside. Sometimes I think about when The Church came, with its words, saints, and sin—and all I can do is cry. The Church set out to save the world from itself and now no one can save themselves from The Church.

There was a book I read once whose protagonist grew up on a Holy Roller commune in the bayou. Her grandfather was a preacher who spoke in tongues and told the children that when the rapture came, those that were left behind would be plagued. He promised these young believers that when they sought their parents' presence, "they'd find only their clothing left in a pile where'd they'd last been." Food would rot, locusts would swarm crops, and blood would run from faucets where water had once flowed. The author of the book said in an interview that they'd written it based on personal experiences that they'd had as a child.

When I'm gone, they won't remember me for tormenting babies with stories written by biblical boogeymen. If they catch me singing gospel songs from the depths of my belly I'll tell them that our country was built by slaves and when they were singing to god, they weren't singing to Christ . . . they were singing to *their* Creator. We'll hold hands and I'll tell them that at every funeral I attend on the Rez, someone asks me to sing "Sweet Beulah Land" and I can never say no. It's not Christ that makes us cry for a place we've never been, but it's the only song the colonizers gave us to help us miss what's moved on to the spirit world.

By the time I'm an ancestor, I will have reclaimed my language and the generations that come to be after I am no longer will sing to their Creator too. They won't sing only when they're in pain or being oppressed. They'll sing because I will have taught them gratitude and the medicine that is giving thanks in song. We'll sing in our traditional tongue and when I am there to welcome them to the place where our ancestors go, they will greet me in languages that were almost lost.

I don't want to just be there for my people after this life. I want them to know me in this one.

One time, I was sitting at the kitchen table peeling potatoes with my mom and grandma, bursting at the seams to tell them a secret, but I didn't know how. After Grandma died, I found out that she'd carried the same secret I had, and that my mom shared in our experience, and bore the same burden of bearing our truth. If there's one thing I can commit to, it's to being remembered as the person who talked about sexual violence. I'll have let my people know that growing up, 1 in 2.5 Indigenous women would be sexually assaulted in their lifetimes, but that we changed that when we taught our children about consent. They'd know by the time I'm gone that no means no means no means no. We would have learned to treat each other as humans again, and would stop having to have conversations about intergenerational trauma once truth crawls out of the wood.

When they tell my story, I want to be the person that young people looked up to. I've worked for rock stars, actors, chefs, dignitaries, grandmothers, and politicians, and have only been treated with kindness by those who defined themselves by their humanity first. I won't be the ancestor who used people like rungs on a ladder, or treated young folks like minds to be mined. I'll have taught that everyone exists within the circle. No one above another, no one more important, everyone seen . . . each feeling valid. I won't be the person who becomes so mighty that she's forgotten to pass the mic or forgets that you can't carry the torch forever. I'll have led by example in such a way that I inspire leaders that want to teach others how to lead.

Every night before bed when my grandma was sick, I'd get on my hands and knees to beg for divine intervention. I didn't care if it was the Creator, Jesus Christ, Allah, or the spirit of Whitney Houston that came to restore her health, I just wanted healing to wash over her. I asked her if she was afraid to die, and she said that she wasn't and that she had a lot of peace. She was ready, though I certainly wasn't, and I realized I didn't even know my grandma's favorite color. What was her favorite smell? When did she first fall in love? These were all questions I couldn't answer.

Weeks passed and as she grew thinner, things I'd never thought to ask began piling up in my mind. By the time I had the courage to share any of them, she barely had control of her motor skills. I knew I wasn't going to be able to have her tell me about growing up during a world war, or whether or not she'd ever regretted giving her life to god or voting for Reagan. I sang to her the song I'd always been asked to sing, telling her about an eternal home called Beulah Land, before crawling into bed beside her the way I'd always done. One of the last things she ever told me was, "Purple." That had been her favorite color.

In the years since her passing, it has become a moral imperative of mine to document the teachings of my elders. Stacks of notebooks and highlighted, dog-eared pages of books line my shelves, filled with their essence and memory. When I am missing them most, I surround myself with the words of my foremothers for guidance and comfort. I want to be the kind of ancestor that, even in my passing, can sit on a shelf, or be present in a book and bring spirit to a hurting heart. A rapper once said, "Writers never die." I'll be the ancestor that writes so that my people can live.

Speaking of life and living it—at this point in time, Indigenous youth commit suicide at three times the national average rate. I know because I've read the studies, but also from personal experience. My suicide attempt happened at age eleven, and my mother's mother, whom I called mom, came to live with us. It didn't occur to me then that she was there to make sure I didn't try to take my life again. Each morning I'd wake to the smell of her coffee and at night I'd watch as she'd meticulously set her hair in curlers before going to bed.

When I was well enough for her to go home, back to our reservation, I had a hard time coping. So much so that my mother took time off from work and drove with me across the country to the foothills of the Rocky Mountains. When I got there, I felt the roots reaching out from the soles of my feet, and grabbing hold of something deep within the earth. It was decided that I'd seek healing with the help of traditional medicine, ceremony, and the support of my grandmother. Years later, she asked me why I thought that suicide had been the answer. I explained that I'd felt so isolated growing up away from our people, knowing I was Indigenous, but never really being sure I could be Indigenous *enough* that far away from home.

She told me that our migration routes existed far before colonial borders and that wherever I was on Turtle Island, I would always be home because I was Indigenous to this land. This was stolen land, yes, but it was our land nevertheless. I felt so empowered by this, and began voraciously devouring traditional knowledge with this newfound lens to look through. As an ancestor, I'll have gifted future generations with the ability to love themselves enough to live, and will have left them with reminders of ancestral rights that help them to stay grounded.

Each day when I left for school as a child my mother would ask me, "Who are you?" I hated that question and the word "hate" wasn't allowed in our house, so I *loathed* it. In retrospect, I am grateful to my mother. My answer was a testament to those I represented, and those I'd leave behind at some point in my life. When I said, "I'm Eryn Marie Wise, Jicarilla Apache, Laguna Pueblo, Spanish, and Irish. I come from the Ollero and Roadrunner clans," I wasn't simply stating fact, I was prayer coming to life, and my mother taught me this. My very existence was thanks to the countless acts of resistance that those who came before me suffered and died for over centuries—believing that one day I'd come to be.

It is deeply moving to know that though you have tried to take your life, or watched those you've loved live and leave theirs, that your very breath has been afforded to you by your ancestors. There are so many teachings and lessons I want to leave behind when I make my journey, and if they ever sing honor songs for me, I want to have earned them. I suspect that I won't be the kind of ancestor that I expect myself to be, I can only pray that I continue to walk in beauty while I'm here. I pray that I leave laughter and hope in the hearts of the lives I'm able to touch, and the determination to live so that others may be afforded life.

ESSAY

Of Land and Legacy

Lindsey Lunsford

Daughters and Sons I Do Not Yet Know,
    I aspire to be the kind of ancestor that leaves something of value behind.
    Land. Your seat. Where you'll stand, and if need be, where you are willing to fall. It is our ancestors who leave us our land, our seat, our place in this world. I aspire to be the type of ancestor that leaves behind a place upon which my descendants can set themselves. That is what my grandparents did for me. And it was no small feat.
    I am typing this essay in my grandmother's home; her pearl of price, her Beulah land. She passed seven years ago, but her home still stands and it is this place that I inhabit, as a young woman seeking to make her claim in the world. For my grandmother, growing up a Black female bastard child in the South during the early 1930s could not have been easy. Not when her would-be fair skin wasn't fair enough to blend in with the rest of her family's, all courtesy of her father. A man who could not deliver her his last name but was able to give her some hints of his mahogany skin tone, just enough to leave her cast away from a color-struck society. So she worked, she studied, and she made way for herself.
    Eventually attending and graduating from Tuskegee Institute (University) in 1947, the same year she married Emmett Lunsford. Emmett was a man a few years her senior and whose velvet skin was many shades darker than hers, a man who accepted and embraced my grandmother. Emmett was determined. His name was Emmett but he spent four years being called "George" as he worked as a sleeping car porter and was a card-carrying member of the Brotherhood of Sleeping Car Porters, the first major labor union to be run by

Black people. Black sleeping car porters attended the trains and were all called "George" after George Pullman, the owner of the Pullman Standard Car Company. It didn't matter their individual age or what their real names were, these men were to be no more than a smile, a wave, and a hand to carry your luggage. Understand, it was honor to have the occupation, no matter how demeaning. It was my grandfather's place to find pride in his position, to find meaning from within the very midst of misery.

Emmett worked as a sleeping car porter, dealing with the blows of segregation and blatant racism so he could pay for his education at Tuskegee. While in school he continued to work, even sweeping George Washington Carver's floors for his work study position. Emmett did this process of alternating between working as a porter and then attending school for twelve years before obtaining his degree and becoming an educator. He would then move my grandmother above the Mason-Dixon line to Indiana, the place of my birth. It was here in Indiana my grandfather would make history as the first Black guidance counselor in the entire state. But that wasn't easy either. To do this, my grandfather had to obtain a graduate degree in the 1950s from a predominately white institution. That's not easy now, but it was unfathomably harder then.

My grandfather could work hard, he could study long, but no matter how hard he exerted himself, he could not get his skin to transform into the shade of white needed to excel in the academic environment he was placed. My grandfather would turn in papers always to have them returned failed and dissected, until the day a white classmate and he tried an experiment: they switched papers. My grandfather turned in his white classmate's and his white classmate turned in his. Of course my grandfather swiftly received another failing mark, but for the first time he got to witness his work, under another name, returned with passing colors. My father advocated to administration and his teacher was forced to pass my grandfather despite his racist proclivities. So, no, my dear ones, don't take your education for granted when your mind comes from those they denied, those they despised, and those they were incapable of ever truly suppressing.

When War World II had called, my grandfather went to Germany, only to learn racism wasn't homegrown but a global occurrence. As he stood on the beaches of the town he was sent to liberate, women would run up to him and his fellow Black soldiers and ask to see their tails. They were told Black men had tails by the white American GIs. All monkeys have a tail, right? Never again could he question the fact that his humanity was not apparent to all. To some, to many, he and his were not human. But it mattered not what the world saw him as compared to what he saw for himself in this world. Emmett

Lunsford was a man with a strong back no matter how many times he'd been asked or forced to stoop; defeat was not to be passed down in his genes. He and my grandmother became the parents of four children. All attended and graduated from Tuskegee University. It is from here that my grandparents Emmett and Frances Lunsford were able to make way, to secure land and to secure their place on this world. No matter where they began in life, they were able to come together and secure a means to start from for their children, for their descendants. In leaving behind physical holdings like land and homes, my grandparents were able to leave behind a legacy. The story of how they came to be, of who they were, was handed down with their descendants.

But there were so many, countless, that were not as fortunate. There are those that by no will of their own could leave nothing but a biological connection to their descendants. Those that could only leave behind in the DNA of a distant child their hair color, their height, or their skin tone to show the world that they were here too. We must never forget those who left their descendants no land because they were hanged trying to protect it. This world will never be able to tell us how many people were forcibly removed from their land. Yet, still, there were those that could not even leave behind so much as their last name because they were torn from their families; sold on auction blocks. Does their blood not carry on in some child's veins?

Since the time of my grandmothers to mine, millions of acres of land were stolen from Black farmers. The twentieth century has witnessed the decline of 98 percent of all Black-owned farmland.[1] It was predicted that by the year 2000 Black farmers would be extinct. If Black farmers were polar bears or tree frogs they'd be put on a list and protected to the utmost. Their land, their customs, their offspring, and their lives would all be protected. They'd matter. But the reality, as told by the numbers and the millions of Black people living in rural and urban poverty, is that their lives only matter to them and their descendants, to those that are related to the struggle. This loss, this evaporation, this hemorrhage of society was not by accident, my distant child. It was by design.

Understand full well that "extensive land ownership among a people and the intelligent use of the land to enhance the income of the owner(s) are basic to the development of a people."[2] We were intentionally underdeveloped at fear of what our brilliance might bring: power. Our land was taken from us. Ninety-eight percent of our farmland ripped away in less than a century. That's a misery that cannot be comprehended by the most empathetic of hearts. Untold children displaced from their inheritance, their birthrights rendered null. How many were forced to migrate to cities that did not love them? To lands they could not claim? It was intentional. Displacement is a tool of control.

They've always tried to control. So to my descendants must I advise: remain a skeptic. In this life you will come to find joy, but never forget the struggle of those that looked like us.

No matter how you look, we are in you. There are multitudes of fair-skinned folks that claim not their great-great grandparents because their very existence was the cause of shame, like my grandmother's father. There are those still whiter with roots much blacker, that will only acknowledge their 4 percent African ancestry over fancy dinner discussions, if at all. So these predecessors, like all of Sally Hemmings' children, are hidden, the Black apples plucked from the white trees.

Ancestor of mine, no matter how far you are from me in time, know we'll never be beyond those that were bound. The enslaved, the sharecroppers, the stowaways, the marooned. The newly freed and the never found. See even now, on whatever day these words find you, understand that we are not free as they would have you believe. We are still prisoners of politics. It is hard for us still, in this time of modernity and post-racialism. Children that will come from us, do not fall for the hype. They lynch, hang, beat, rape, and kill us still. They still enslave us in prison sentences, abject poverty, and wanton stress. We preserve as we always have but know what's in you. It is what's in us that will let you come to be, despite the attacks they enact on those that look like us, like you. If only our locks were just dreaded and not feared? If our booming voices, curvaceous figures, and unstoppable minds weren't such a genesis for envy and extermination.

I know not the burdens that you'll face, only that I do not live in a post-racial society and nor will you, no matter how far away you live from me in time. I tell you this not to cause you fear, only to spare you untruth. For lies and silence be the greatest barriers to understanding and connecting to our ancestors.

My child, as the concerns for the environment grow, know that you, we, we are not the cause. Soon they'll ask it more than they do now, "Who ruined the earth?" It was *them, they*, those that oppress. You must never forget that Black hands never hurt the Earth. We fed her, even under lash and heat we shed our tears and hopes into her and we tended her. No my child, the world was consumed by the consumers, those that ate the land. Those that ate our names. Destruction is not our legacy; it holds no place in our lineage. Hold your head high when the wolves howl, for your people did not cause this world to twist in the ways it has. It was *them, they*, those that oppress. You know them. They wear suits, they speak in numbers, and their eyes cannot detect light. They don't love, they exploit, indenture, and promote fester. They are the true masters of commerce, the captains of industry, and they tried to do us all in.

At this very moment they are working to ensure that you do not exist. That fewer and fewer Black babies are born. They call it population control and it is happening at the highest levels of government and order. It is the goal of the new millennium to see the rapid voluntary infertilization of Sub-Saharan African women realized.[3] Resist this. *Resist it.* Poor Black children have as much right to be here as anything or anyone else, never forget it.

Children of tomorrow, if there is anything I don't want to leave you it is my pain, but I know it's to be transferred. As much as I dread it I, hidden in your DNA will be the epi-memory of our internment. As my grandmother's tears and trials were sewn in her blood meal, so will mine be shared with you. For that is how we learn to adapt, my pain will be wrought unto you, so that you may have a head start, learn what to avoid and what to run from. It is how we continue.

But you'll also inherit my magic. The recipes for your redemption. The ability to survive, to even thrive, when the odds are stacked against you. Our connection: I'll whisper to you as my ancestors whispered to me. We, a people who've faced disconnection for centuries, we can never be fully separated from one another, for your ancestors and our spirit reside in the same plain. We are then; we are now; we are tomorrow. We are omnipresent: we are linked.

Let it be known that our greatest aspiration is to pass to you our land, a place to sit in this world. Then comes the need to pass to you our legacy, so you know who you are and what you are composed of. But if fate robs you, as it has countless, of the ability to inherit your land or your legacy. If no one sings to you our family's song, if it is to be that one day you find yourself without me; without stories of where you came from, of who you are—know that *some things can never be lost.* Even without your name, you'll always be what you're made of. What we made you of. You have the heart of them field-working mamas, the pride of those sleeping car porters. And that beautiful face of yours, well it tells stories that you'll never hear. Of those that came before us. The ones that traveled, that overcame, that loved, our ancestors.

REFERENCES

R. Grant Seals, Wimberley, R. C., and Morris, L. V. 1998. *Disparity*. Vantage Press.

Sachs, Jeffrey. 2015. *The Age of Sustainable Development*. New York: Columbia University Press.

Vallianatos, E. 2012, September 10. America: Becoming a Land without Farmers. Retrieved July 17, 2018, from https://www.independentsciencenews.org/environment/america-becoming-a-land-without-farmers/

ESSAY

Cheddar Man

Brooke Williams

Recently, Cheddar Man "flirted"[1] with me. This was not the first time. I don't mean that he has romantic feelings for me, but rather that news of recent discoveries in the DNA of this 10,000-year-old man grabbed my attention and will not let it go.

I've learned over the years not to ignore these flirts and instead sense that they signal a message coming from my inner, unconscious world. Learning to truly pay attention to what attracts my attention has added an important dimension to my life.

Cheddar Man is not his real name[2] but was given to him when he was discovered, curled beneath a stalactite in a cave in England's Cheddar Gorge in 1903. Years later, using newly developed techniques to extract and learn from ancient DNA, scientists began piecing together Cheddar Man's genetic heritage. In 1997, newspapers across England featured a photo of a local schoolteacher who, based on recent analysis of his mitochondrial DNA, was confirmed to be Cheddar Man's direct descendent.

In 2001, I was traveling around England with two purposes: to find out more about my ancestors for a book I was writing; and to explore as many prehistoric sites as possible. Reading a book on Cheddar Man that I'd discovered in a local bookstore, I realized that the ancestors with whom I'd grown familiar didn't just drop from the sky, but that they had ancestors who had ancestors *who had ancestors* going back to the first life appearing in that steamy swamp 3.8 billion years ago. Cheddar Man combined the two reasons I'd traveled to England by hinting that I might have ancestors who helped tilt *menhirs*—the "standing stones" of Avebury and Stonehenge—into place; or hoist the capstone at Trethevy Quoit.

Cheddar Man's most recent flirt came from an article in the Guardian[3] announcing the results of new tests on his DNA. During the years between flirts, the scientific knowledge regarding Cheddar Man's DNA had continued to increase. Geneticists from the British Museum had found viable DNA in powder extracted from the small hole they'd drilled into Cheddar Man's forehead. From it, they determined that Cheddar Man was a Western European hunter-gatherer living during the Mesolithic era, the five-thousand-year span between the Pleistocene and Neolithic periods (11,000 to 6,000 years ago). In addition to revealing the time and place in which Cheddar Man lived, his DNA revealed that he had blue eyes, black hair, and dark skin.

Dozens of magazines and newspapers announced the latest Cheddar Man findings. Many included a photo reenacting the discovery of Cheddar Man's bones as they were found laid out in the cave in 1906. He is on his right side, his legs bent. A photo of Cheddar Man's head, recreated by paleo-artists, accompanied many of the articles. He's looking up to his left, and although the corners of his mouth are turned slightly down, he is not frowning, as there is an unmistakable gleam in his eye.

Growing up, ancestors played a minor role in my life. We were told only of those dead relatives with admirable and faith-promoting stories. I don't recall being encouraged to imagine that my ancestors might be alive in some form or playing an active role in our lives. I had heard or read about people of different cultures—Chinese, Japanese—who worshipped their ancestors, and that modern Mexicans contacted their ancestors yearly during the Day of the Dead. However, we were taught that we would meet "those who have gone on before us" only at the moment of our deaths. None of this seemed practical (or interesting), and I didn't think much about ancestors until seeing the movie *Amistad*. This film tells the true story of the takeover of a ship, bound for America, by a group of Africans who were to be sold into slavery upon arrival. There is a scene that I've not forgotten during the decades that have passed since seeing that movie: The Africans are on trial. One of them, Cinque, is about to testify. He is speaking to his lawyer, John Quincy Adams.

CINQUE: We won't be going in there alone.
JOHN QUINCY ADAMS: Alone? Indeed not. We have right at our side. We have righteousness at our side. We have Mr. Baldwin over there.
CINQUE: I meant my ancestors. I will call into the past, far back to the beginning of time, and beg them to come and help me at the judgment. I will reach back and draw them into me. And they must come, for at this moment, I am the whole reason they have existed at all.

I will never forget the feeling of wonder I had, imagining that all of my ancestors might be out there somewhere in other dimensions, in other worlds, waiting for me to "call into the past" for help.

While I haven't called out verbally to my ancestors for help in the way that Cinque did, I have called out in my heart—over and over since the arrival of America's most recent president. While Trump seems intent on destroying American democracy while making a mockery of it, loosing the corporate hounds to ravish anything sacred and putting more money into the pockets of his billionaire friends—as well as his own—while taking it from the rest of us, his main accomplishment has been to rip the lid off the caldron in which white male anger has been smoldering for decades. (Trump's other "accomplishments" to date consist mostly of undoing anything that his predecessor President Obama, America's first Black president, set in motion.) Since his election, Trump has tried to ignore, failed to condemn, and, at times, cheered on the perpetrators of racial, anti-Semitic, and sexual violence. If forced to name one positive aspect of the Trump presidency, it would be that he has exposed the fact that America is clearly a racist, misogynistic country. He's opened the caldron that, left sealed, might have exploded.

I'm a biologist by education. I write about modern humans as biological organisms still subject to the forces and power and beauty of evolution. Race, I was thrilled to learn decades ago, is a very recent phenomenon in context with our deep history. Why has it become the root of some of our deepest problems?

This was the question I was asking during the weeks leading up to Cheddar Man's most recent flirt. It's as if he's been floating out there with all the ever-dead people, worried about the living (as I believe all dead people spend their time). Then suddenly, there he is—his dark face, his twisty black hair, his blue eyes—in a hundred articles spread across the internet. Unavoidable. I read every one.

Like me, many white Americans have ancestors that came from Europe, and specifically, England. Cheddar Man's geneticists (we all should have personal geneticists giving us annual checkups—at least—to remind us who we really are and where we came from) say that 10 percent of the genetic material in modern Brits comes from these ancient hunter-gatherers.

My deep history is complicated. With so much interbreeding—Sapiens with Neanderthals, farmers with hunters—and so much movement (not to mention how history is filtered through different belief systems), opinions vary. Here, based on my best interpretation of available information, is the timeline for my personal history.

Two hundred thousand years ago (7,000 generations) our species, *Homo sapiens*, first appeared in Africa.

Forty thousand years ago (1,500 generations), *Homo sapiens* first arrived in England. The Ice Age was in full force; the seas were low; and England was connected to mainland Europe.

Twelve thousand five hundred years ago (500 generations), Cheddar Man's ancestors came into England from the Middle East.

Ten thousand years ago (300 generations) Cheddar Man was born in England.

Six thousand years ago (200 generations) farmers from the Middle East came to England.

Two hundred ten years ago (6 generations) my great-great-great-grandfather was born in Shrewsbury, England in 1808. (He came to America in 1863.)

While all these milestones are significant elements of both personal and collective history, the introduction of farming in Europe is of particular importance. Much has been written about the transition (many refer to it as a revolution) our ancestors made from hunting and gathering to farming—as we moved from the Mesolithic to the Neolithic period. Farming required waiting. (I wonder if those early farmers suffered from wanderlust. After thousands of generations as nomads, did their cells scream to move?)

From this waiting came modern life as we know it. Besides the ability to control our food, it also allowed for permanent architecture, increased fertility, the ability to feed more people from a smaller land base, and specialized craftsmanship leading to innovation and technological advances. While these can be viewed as accomplishments, others see the Neolithic in a less positive light. Jared Diamond calls our adoption of agriculture, "the worst mistake in the history of the human race." Runaway technological development, gender and race inequality, property ownership, and degenerative disease all resulted from adopting the sedentary life of farmers.

Now that Cheddar Man, our ancestor from three hundred generations ago, has come forward with his dark skin and his wavy black hair, I need to ask: "Why did our early farming ancestors—from two hundred generations ago—have light skin?"

Dark skin was an advantage while living in the intense equatorial sun where we came from. The stone knives and bone harpoons found with Cheddar Man suggest a varied diet of meat and fish, a diet loaded with Vitamin D. In contrast, the diet of early farmers was based on two or three domesticated grains. Discovered remains offer clues that they suffered from soft bones, various deformations, and other problems associated with a dearth of Vitamin D (which is created underneath our skin and supplemented by diet). Moving north to areas of less intense sunlight exacerbated this problem. Our farmer ancestors adapted to these changes by evolving light skin.

We evolved light skin, it seems, from weakness not strength. Not much to build white supremacy on, is it?

Where does this supremacy come from, then?

We know that light skin is associated with both white supremacy and the advent of agriculture. We also know that innovation and technology flourished in places where agriculture had been established. There are many theories as to why this is not a coincidence. Two of the most common are that the work of a few farmers fed many others who were then free to think and create; and being stationary and no longer nomadic allowed us suddenly to accumulate more than we could carry at any one time.

I have another theory that links agriculture, technology, and white supremacy. As noted above, archeologists studying our early ancestors' remains can tell the farmers by their deformed and fragile bones. Evolving lighter skin was one adaptation to this. Evolution, remember, is simply adapting to changing conditions. Technology in the form of guns has been a different kind of adaptation to this weakness. Moreover, gun control was initially a series of statutes instituted by (white) lawmakers, fearing revolt, which barred (Black) slaves from having guns or other weapons.

Our white skin evolved not out of strength but weakness. Let that sink in for a minute. Perhaps this explains why "my people"—privileged white men—will do anything for power, and why, when tenuously hanging onto their last vestige of power, they will change the rules, make new laws, sign executive orders reversing anything that does not enhance their power (over those of other races, other genders, or other belief systems). This weakness may explain what seems otherwise unexplainable. And what makes the least sense these days? Guns. It seems to me to be the ultimate in fear and paranoia that just about anyone can purchase an assault weapon designed to kill as many people as quickly as possible,

White skin evolved due to a weakness not a strength.

For some, the Bible explains it. Recall that after killing his brother Abel, Cain received a mark from God. One interpretation is that this mark set Cain apart as a wanderer and fugitive, intended to warn others while protecting him. Another (many feel misguided) interpretation of the story is that the mark is actually "the curse" of black skin, which was passed down through the ages and used to justify the slave trade and, later, segregation. However, there is nothing in the Bible suggesting this is even a remote possibility.

God is also given credit (or blame) for "manifest destiny." Columbus "discovering" America was part of the early efforts of Spanish and French monarchs, who ordered exploration of the "New World" as part of their divinely inspired mission to convert natives to Christianity, while plundering the riches found there. As historian Donald M. Scott describes, manifest destiny "was also clearly a racial doctrine of white supremacy that granted no Native American or non-white claims to any permanent possession of the lands on the North American continent and justified white American expropriation of Indian lands."[4]

The mark of Cain and manifest destiny may be at the root of the problem Angry White Men have with the rest of the world, helping to explain the hypocrisy, illogical behavior, and violence toward the other we're experiencing today, as white supremacy once again surges in what we can only hope is its last desperate gasp.

Throughout history, a belief has persisted that somehow a god anointed weak white farmers and their descendants as special, superior, *exceptional*. Furthermore, this myth has perpetrated ill-gotten power, influence, and wealth (*greed*) by suppression and control of any perceived threat. Importantly, this suppression has been made possible not by higher intelligence or superior creativity, but by violent force enabled by more effective and powerful weapons. This weapon technology originated in the minds of European designers and craftsmen, their creativity unleashed as they no longer had to find and kill their own food.

A god favoring one weak race that only maintains its stature through force and physical violence, is not a god that I'm able to imagine.

Evolutionarily, white superiority is based not on strength, but a weakness for which better weapons has been a terrifying adaptation.

This morning, in and out of sleep, the photos I'd seen of Cheddar Man came back to me. First, I saw the one of his body of bones as they'd been found in the cave. In my altered sleep-state, his bones were transparent and, in them, glowing blue liquid began flowing as if lit somehow from within. First his lower legs began to glow blue. Then the liquid flowed into his femur and pelvis and then up his back. I waited as the glowing blue liquid seeped slowly to the tips of each rib while simultaneously moving up his spine and into his skull where it began boiling. The blue liquid filled his collar and shoulder bones and into his arms and hands and, when the tips of each finger began to glow, I realized that Cheddar Man had come alive in my unconscious. He had come from the dead to remind me that we are all one people; he came to remind me of my evolutionary responsibilities. He came to inspire me in the ways of our deepest past because they are the key to our longest future, as they have always been.

Then, in my early morning mind, the photo of Cheddar Man's recreated head appears. At first, I assume this photo is identical to that which has accompanied every recent article—of the man flirting with me. Then he turns his blue eyes, which had been staring up and left, to meet mine; they still have a gleam, as if he's waiting for a sign that I've received what he's come back to give.

ESSAY

Formidable

Kathleen Dean Moore

When my big sister burst into the Trailside Museum, my father was giving a talk on the subject of canine teeth. He held a raccoon skull in his hands. His pet raccoon sat on his shoulder, peering intently at the skull.

"Kathy fell into the river," my sister shouted.

My father dropped the skull, thrust the raccoon at the nearest person, and ran for the door.

While he skids on the icy path and leaps over snowbanks, I can tell you that I didn't exactly fall into the river. I jumped in. I was walking along the stone wall that edges Rocky River, and there below me, on a litter of snow and sticks and fallen leaves, was a red ball. I wanted it. So I jumped off the wall. And dropped like a shot through the floating leaves. As I flailed around in bubbles and ice, I found a willow sapling somehow, and pulled myself high enough to breathe. But then I was stuck.

So here's what I remember, although it has been countless decades, and all the details of that day have been told and retold, the story probably mashed in the making. I remember that the current bent the willow and pulled at my coat. I was scrabbling my feet against the wands. I remember my father's rubber boots sidestepping along the slender trunk of a maple that had toppled into the river. He moved step-by-step, holding on branch by branch. The trunk sank under his weight, and water slid over his boots. I remember that water—smooth as silver. He bellowed at me to let go of the willow and reach for him.

Back at the museum, he was furious. He tried to yell—*Who would jump into a river in January?*—but he cried, and what man can scold when he is crying? My mother was the one who was mean, making me strip to my underwear and sit on top of a ladder next to a heater vent mounted to the ceiling. Can you imagine the mortification of a nine-year-old girl crouched on a ladder, half-naked, half-drowned, in front of a couple dozen people wearing binoculars and ear-flapped hats, *everybody* wondering what kind of idiot would jump into a river in January?

But maybe not an idiot. Maybe just an ordinary person who wanted a red ball and did not understand that the ground would give way under her feet.

*

So if you ask me what kind of ancestor I want to be, I will tell you: like that formidable father—that angry and overcome by terrified love, that brave in defense of what is young and endangered.

The small lives on the planet, present and future, are in a "to be or not to be" moment, an emergency of our culture's own making. And aren't you and I, all of us, called to be that father, that mother, all of us called to save the young ones—the future generations of the human species and a myriad other species on the beloved planet—from the wreck and plunder of a fossil-fuel-driven, infinitely expansive, dangerously reckless way of life?

People in the wealthy north just want that red ball, that half-ton pickup, the Kobe steak, the propane BBQ, a third baby, that pretty pistol or iPad. It's music on demand, movies on demand, everything we want on demand, guaranteed two-day delivery of anything in the world—all of it powered by fossil fuels, powered in turn by a radical capitalism that knows few legal or moral constraints.

That way of life is easy to get used to, as long as we never understand the cost, as long as we never look back and see the ruined forests and souring oceans, the poisoned estuaries, the deaths that are required to produce these pleasures; as long as we never look forward and foresee the ecosystem collapse, the dead zones, the melting ice, the dangerously changing currents of sea and sky brought on by global warming.

We might think the ground is solid under our feet. But what looks like solid ground is a pretty layer of leaves and sticks floating on a growing void. Any culture that prides itself on accumulating Earth's wealth instead of replenishing it, any culture that gobbles up the fecundity of the planet instead of nurturing it, any economy of infinite extraction, will empty its own foundations.

One after another, scientists burst into the room and shout a warning: "Unless all nations take immediate action, by the time today's children are middle-aged, the life-support systems of the Earth will be irretrievably damaged."[1]

Unless we stop fossil fuels, the small ones will live and die in a world of violent chaotic weather, spreading disease, water shortage, collapsed agricultural and fisheries systems, wars for resources, and massive movements of people driven from their homes by wildfire or flood.

It's the children who will be swept away, the young ones of all species, and their children after them.

This is a betrayal of our love for them.

The grown-ups on this planet have an absolute duty to save the children from global collapse. "Because we swore and vowed to every god we ever imagined or invented or dimly sensed that we would care for the children with every iota of our energy when they came to us miraculously from the sea of the stars," poet Brian Doyle wrote. "Because they are the very definition of innocent, and every single blow and shout and shiver of fear that rains down on them is utterly undeserved and unfair and unwarranted. Because we used to be them, and we remember, dimly, what it was like to be small and frightened and confused." And I would add, because we held them in our arms and said, "I will always love you. I will keep you safe. I will give you the world." We promised them.

*

So what response will match the scale of the moral and mortal peril? Who do we grown-ups need to be? It's hard to say. My friends tell me to take it easy. *Stop being an alarmist. It does no good to call names. And if you care about the kids so much, how about you drive them to soccer practice this week?*

But shouldn't we all act with an intensity that matches the danger of the moment? This is a real question; it haunts me. Don't I have a duty to sound the alarm when the danger is real? Surely there is no such thing as an alarmist when the storm-surge, the fire-storm, the storm-wind, reach the doorsill. Silence in the face of mortal threats to what I most deeply value violates my own moral convictions. How can I claim to love the children, if I don't name and condemn and fight against the system that empowers those who would destroy them?

*

Years ago, I watched a raccoon defend her babies. She was formidable. What I remember: an angry dog barking. In the yard light, a ringed tail, an arched and bristling back, lowered head, flash of white on bared teeth. I remember the bark and screech, the hiss, a yelp. Tirelessly lunging and swirling, the raccoon *would not* let the dog get behind her, where three babies were staring from a shed in uncomprehending wonder, unimaginably cute and stupid. In the face of those teeth, the dog wandered away.

We humans, like the raccoon, are born to defend our young. This is why we have evolved the particular bulge of our brain and the cleverness of our hands and the closeness of our communities, this is why grownups hoe the corn and commute to the office: that our offspring will live to bear their own children. This is foundational human nature—this courageous, ferociously protective love of the little ones.

So what accounts for the general listlessness in the face of global warming's grave threat to future generations? Listlessness is something that needs to be cultivated. And cultivated it is, by the fossil-fuel industries and their political minions, with breath-taking levels of funding, skill, and determination. Their "game" plan: Convince people that global warming is not a real threat. Or if it is, it's a threat that we can adapt to, with all our technological skill. Direct the worst effects of global warming toward those who have no voice to protest—plants and animals, future generations, and those who are silenced by racism or economic desperation. Buy politicians and agency heads who will gut the regulations and silence the science. And most effective, use a relentless media assault to change people's understanding of who they really are: not ancestors, but heirs; not moral agents, but consumers; not community members, but competitors; not creatures of the Earth, but miners of its riches; not powerful actors, but victims of inconsistent and autocratic cruelty; not free, but confined by constant distractions. *Look, here, such a pretty red ball.*

A spokesman for the American Petroleum Institute once leaned over me and said, "Don't you ever, ever, ever, ever underestimate the power of the fossil fuel industry." I assured him that I would not. That said, I know this for sure: that if there is a force on the face of the Earth more powerful than oil, it is the ferocious, protective, blindly brave love for the children.

*

Last winter, I walked the snowy McKenzie River trail with my little grandson, holding on to his mitten to keep him from skidding into the creek. He laughed and tugged, toddling for the water—of course. I scooped him up, brushed the snow off his forehead and kissed it, and I thought, as I often do: This love is so big, so intense, I feel it radiating, like a sort of heat energy, from my body. I thought, if this is true for me, it must be true in some way for every other parent or grandparent or aunt or uncle, in every corner of the world, in every house and hut, in every burrow in the boulders and nest in the trees. So if I take what I feel and multiply it by—oh I don't know, fifty billion, a trillion?—then I understand that the world shimmers with love, the incandescent urgency for life on-going.

If anything stands in the way of destruction, it will be this force. This is the force that will give us—the ancestors—the courage to go out on a limb in the dangerous flood, the courage to reach for the children.

INTERVIEW

Caleen Sisk

Brooke Parry Hecht and Christopher (Toby) McLeod

HECHT: *I would love to hear the earliest stories of your ancestors that you carry and how those stories define your purpose.*

SISK: As Winnemem Wintu people, we believe we were put here by the Creator. We didn't come across the Bering Strait. While some people went back-and-forth across the Bering Strait, we did not. Our story says that we came from Buliyum Puyuk, Mount Shasta. We came out of a spring on Mount Shasta as a spirit that took form. Before that, Creator had the spirit beings staying inside a big mountain, with the sacred fire inside that mountain.

Then, Creator came to say that he had created this other place, and he needed helpers to take care of it. He instructed the spirit beings to choose a physical form and take on a job to care for the place that he had made. Spirit beings began to choose what they wanted to be. Hummingbirds went out and declared what they would be doing. Acorn trees—several different varieties of Acorn—went out, and they said they were going to feed the people. Then, the bigger trees like Fir and Pine and Cedar, went out. Then, the animals started choosing what they would be. Elk and Moose, Bear and Deer—all of them chose what they were going to be, and they went out. All of the four-leggeds went out. All of the flyers went out. All of the swimmers went out.

One of the last ones going out was Salmon. And Salmon went out. Then, the last one in there—walking around the fire—was trying to figure out, "What could I be?" He realized that most things were already chosen, and he had waited too long. Finally, he said, "I'm going to be human," and he went out the door and down the stream. Creator thought, "Hmm. That one's going to need a lot of help." So, he called back the Fire Spirit and the Water and Mountain Spirits, and he asked them to take care of that one. This way, the people could always come and pray to the Fire Spirit and get in contact with the good fire inside the mountain—to set them upright again, help them, and protect them. The Water Spirit would give them life, a way of life. The Mountain Spirit would protect them and provide them with all of the food they needed.

Salmon heard what was going on, that Creator was worried about the human. Salmon came back to say he was willing to give up his voice for humans, so that they could communicate. Maybe, humans would have a better chance to do the right thing if they had a voice. So, Salmon gave us that voice. Since that time, Salmon has been a special relative to us in that we are responsible for speaking for Salmon because Salmon gave us their voice.

That's why we have that connection with Salmon now. It is in our stories from long ago, and we have passed that on to our kids. This will go on as long as there are Winnemem people who believe in the Winnemem way of life. That's the struggle we have today, staying true to the Winnemem way of life, regardless of all of the modernization, technology, other religions, and education. But, as long as we can hang onto the Winnemem way of life, we will keep following it.

HECHT: *Could you say more about prayer on the mountain? Since the beginning, you have prayed to the fire spirit, the water, and the mountain. Can you also pray to your ancestors?*

SISK: Yes. Because, the spirit beings are in different forms for different things in your life at different ages or different times of difficulty, or when life changes around you—then there are helpers for us to get through those times.

For example, California is now in a drought, which means we should be in ceremonies about that, communicating with that, instead of trying to control it. Maybe this marks a time that we should be working to better understand our connections with water. Most Californians have lost a close relationship with water. The tribes that continue to do four-day fasts. . . . You think four days is not a long time, but spend four days without water, and you have a new way of looking at it. Maybe, more people should be taught about water that way.

HECHT: *From what I understand, the Run4Salmon is such a teaching—and a response to what is happening with Salmon and water. When you prayed for guidance regarding your responsibility to Salmon, the Run4Salmon ceremony came as a response to that prayer?*

SISK: Right. When you go to the mountain or you go into prayer, then you have to be ready to carry out that prayer. It's not like magic. We have to work for these things. We have to make them happen too. We have to show the Creator that we want this and that we're willing to take the steps to get there.

We've been fighting against the Shasta Dam since the 1940s. We lost the first go-around; they built the dam and flooded 26 miles of our river. Our Salmon lost their homes on that river. And so did we. It stands to reason, because we believe that whatever happens to Salmon, also happens to us.

Since that time, we haven't been allowed to be on the river, other than visiting our sacred sites. But, those are still our sacred sites and still our village sites. Those are still the sacred places that we have to go to if we're going to continue to be Winnemem.

The Salmon are the ones who we need to follow. The Run4Salmon is a reflection of that. Once we understood that, we just had to figure out how we were going to do it. The Run4Salmon can't just be on our river, because the Salmon follow a cycle from ocean to mountain. These Salmon will swim five hundred river miles to get to this mountain, to get to the Buliyum Puyuk, where they came from.

But the Shasta Dam has been blocking them. The Run4Salmon journey, this prayer, is meant to wake people up—everybody. It's not just the Winnemem who are affected by the loss of Salmon, by the loss of living water. Everybody is affected by this. That is one message we took on this prayer.

HECHT: *You have shared some of the oldest stories you carry—about creation, human purpose, and your relationship with Salmon. Do you have Salmon stories from more recent times reflecting the challenges that you have been facing?*

SISK: People think that the Mount Shasta Salmon have been gone for a long time. But, during my grandparents' and my parents' time, our Salmon swam in these rivers. For us, we are just one generation away from these stories, from seeing the Salmon in the river. However, the prayers for our Salmon were set in motion a long, long time ago.

The disappearance of our Salmon started with the US fish hatchery on our river in the 1870s. Our tribe was totally opposed to what they were doing to the Salmon, and so we went into ceremony and performed a war dance. At that time, anyone from the tribe could be killed and turned into cash. All anyone had to do was turn in an ear or a scalp to the justice of the peace. People were paid to kill any age, any gender, of an Indian of any tribe. So, it was a dangerous thing for our tribe to oppose the hatchery, but they did. When the Winnemem did a ceremony asking what to do about the hatchery, they were told to send the Salmon through the ice waterfall on Mount Shasta. So they did. That's the story that has been told over and over. "They sent the Salmon to the ice waterfall."

When we did a war dance on the Shasta Dam in 2004, we learned something important; we learned where the Salmon were waiting. The only place in the world that our Salmon survived was in New Zealand (and the Salmon had been sent many places in the world by the hatchery). Our Salmon still live on the South Island, where they have a mountain, Mount Aoraki, that has ice waterfalls.

We believe that the reason the Salmon are successfully swimming in New Zealand today is that the prayer went with them. That prayer opened that way up. Through the Run4Salmon prayer, it all fits together. It's a full circle prayer.

Our hope is that now we can do what needs to be done to bring our Salmon home to the McCloud River and Mount Shasta. We want to bring our Salmon back from New Zealand and build a fish passage around Shasta Dam. When the Salmon are back in the McCloud River, everything will change. Right now, the trees don't have Salmon. The animals don't have Salmon. Everything will start changing for the better when the Salmon are back. Science is catching up to this fact, but it is still so hard to convince people of this.

MCLEOD: *If our descendants were to look back on this time and see a turning point where things changed for the better, what would it have been, do you think, that would have turned things around from their point of view?*

SISK: It has not happened yet. But I think that if people actually start revering water, and start creating jobs that help water, that would bring back some of the waterways in a way that they were before. Replanting of Tule grasses would be an indication that people understand that water should have Tule grasses and that the Tules do a job for everyone else. Tules purify the water and allow the water to live. When you take all of that away, as the current society has done, you're shipping water that has nothing in it.

After this water is shipped south through the aqueduct, it needs to be revitalized. They even have to put air back into the water because it's come from this dead zone. Water needs riverbeds that are alive. Water running through living riverbeds gets energy, minerals, and oxygen. I don't know why we think we can just shoot water through a pipe and it will still be the same kind of water—that it will be healthy water. The way Creator put water here was so perfect it could exist like that for thousands and thousands of years inside the mountain and still be good. But, people are not gifted like that. Even though we can run that water through a pipe, and we think we control that water, we're actually just running poison water. That's the time we are in now. The towns that have been running water through these pipes since the 1800s are now running poison water, and they can't even use the water anymore. And some people think, "Well, that's just happened to that town. It's not going to happen to us."

When people look back on this time, I hope that my descendants would say that we did everything that we could. From the time that we were given this way of life, the Winnemem were the people singing to the water. And we never stopped singing for that water. If we can teach other people that this needs to be done, regardless of whether that water is coming into my village, or whether we're actually using that water. Creator didn't put us here with a shortage. We are creating that shortage by poisoning the waters that Creator gave us as drinking water.

So, I hope that people in the future will think, "You know, those Indigenous people were on the right track."

MCLEOD: *When we went to Standing Rock, people seemed tuned into the idea that water is life, and water is sacred. That was coming from an Indigenous point of view, and I think that shift from protesting against something to praying for the protection of something was a real cultural moment. This happened because of Indigenous leadership. Could you talk about why it's important to not just simply be in opposition and protest, but to balance that with reverence for water, for the ceremonial?*

SISK: Sharing the idea that all of these people are protectors was the first step. When you're a protector, you're accepting responsibility. I think about Standing Rock in a lot of different ways. Some bad things happened to people there, and the world watched. The protectors paid the price. Part of our work now is trying to teach others about how to be a protector—to have a stand, a purpose, and an understanding of the environment as a whole.

I think that when we figure that out, that the opposition will actually lay down their guns and walk across the river to be on our side—because they would see themselves as protectors too. So, we haven't broken through yet. We haven't hit that tipping point.

I think about the ceremony we conduct every year at Coonrod—the world renewal ceremony. All the tribes here used to have world renewal ceremonies. But in the late '80s, my Grams said, "We're not doing that anymore. We're going to get ready for the change. There's a big change coming. And the world the way that we know it now cannot be renewed like it is anymore. There's a change coming, and we have to get ready for that change. We have to know how to go through that change. A lot of people will suffer and a lot of people will be crying, but people will go on through that change."

Grams was not talking about an army coming or anything like that. It's Mother Nature making this change. She's saying that people are too comfortable. They don't get too cold, they don't get too hot, they don't get too wet. . . . And so when it gets too cold, a lot of them are going to die, because they don't know how to survive in the cold. When it gets too hot, a lot of them are going to die because they don't have the knowledge about how to be in that hot of a condition. For many people, their comfort zone is very small. Indigenous people have very wide existence; their comfort zones have to be wide.

HECHT: *You have said that one of the teachings of Salmon is how they adapt to change. As they swim downriver toward the ocean, Salmon need to undergo dramatic change. I would like to understand more about how this wisdom from Salmon is paralleled by your practices of prayer and ceremony—how you change your practices in response to the changes and challenges you experience.*

SISK: When the young Salmon swim downstream to the San Francisco Bay Delta, it's a saltwater area, and they don't decide, "We made a big mistake. We're not saltwater fish. We better swim back up there, where we are from." We believe that they carry a prayer. They know they are going to change, and they know that the Creator is going to take care of them. That change from freshwater to saltwater can't be easy, being a little fry, but they do it.

Then they swim out to the ocean, and they're out there anywhere from four to seven years. Then there comes a time when they decide they're coming back upriver. They come back to the exact same estuary and, as adult Salmon, they need to change again. They don't say, "This is way too hard. Now, we're adult Salmon. This is the way we are, and we're not going to change." We believe they do change because they are following the Creator's prayer. No matter how hard it is to make that change, they're going to do it, and they stay in the estuary until they do.

After the Salmon acclimate, they start swimming up the river again, and they swim to a particular place. They go to their particular rivers, whether they were born in the Feather River, Butte Creek, or Clair Creek—whatever waterway they are from. The McCloud River fish are the highest on the river run. They have a long way to go and a lot of elevation to gain. These particular fish have to change water texture, they need to be able to swim in aerated water, and they need to be able to jump. These fish are mountain climbers.

Today, the absence of the Salmon and the presence of pollution in the water is like the miner's canary: whatever happens to the fish, whatever happens to the water happens to everything. Nothing can live without water. Somehow, though, this society has gotten away from the indicators, the things that would show us, "We're going the wrong way."

For example, we see that right now, water and fire are way out of balance. And so, we have gone into prayer to seek guidance. In response to prayer, we are building a round house for the fire and water Big Heads—spirit beings who are sacred teachers that come from the stars—to come into the round house to teach us. We know we don't know everything, and so we call on the ones that do.

While we do pray in response to current circumstances, there are some things that we have done, and continue to do, since the beginning. Since the beginning, we have been a tribe who goes back to the doorway we came out of, and we sing for the water. And as soon as we sing for that water, that water comes up out of the ground. It's an artesian spring, so there's no runoff that's making this water. It comes up from way down deep in the mountain, and it comes up very old, fresh, cold water. It comes up, and we sing to it. By the time we're through singing, that water has already passed us by. It has already gone down the stream. We know that it's going out to the ocean, and it's going to go around the world. We sing for water for the world. It's going to come back as snow. It's going to come back as rain. It's going to come back into the ground.

It's going to come back as happy water, because we sang to it. We just hope that other people are doing the same. But not enough people know how to do that or know that they should be doing that. Water is so important to every living thing, but it is so taken for granted. It's thought of as a commodity.

HECHT: *You are saying that more people need to sing to the water like that. . . . There are so many of us who are cut off from our ancestors and the land practices that were closely related to the homelands and homewaters of our ancestors. Many of us don't even carry stories from our ancestors. I believe you have called this, "being asleep." How do we wake up? What thoughts can you share for those who have lost this connection and are trying to find their way?*

SISK: Yes. That's a tough question, because people should be able to find that answer in their own religions. But when there is nothing there, people are creating new religions. Mount Shasta is one of the most significant spiritual mountains in the world. It's a draw for people who are in search of some spirituality, or some religious practice, or some way that they can feel better.

Just as people are seeking these spiritual places, others are offering, "You can pay me six hundred dollars, and I'll take you to that sacred place, and we'll do prayers." We fight with the Forest Service all the time to protect Mount Shasta, but other people's rights, or public rights, allow other people to do whatever they want there. Some people have turned it into a money-making thing. They'll go to a tribe, and they'll learn some things. Then, they'll go off and hold their own ceremonies or hold their own gatherings and take people on journeys to these sacred places. You can find it online everywhere. I'm not wanting to change what people believe in, but I think this is a problem.

For Winnemem people, Mount Shasta is so sacred that we are only supposed to go up there once a year—and only when we're old enough (fourteen to fifteen years old). No children are allowed to go there. We need to be old enough to know how to be there, to understand the relationships that are meant to exist between people and that place. It's that powerful.

For example, some of the grasses and other plants there take fifty years to root and five hundred years to become adults. Some spiritual seekers go there and sit on the grasses because they want to experience the sacredness of this place. But, when you sit on the meadow, you're almost against the sacred. You're ruining the sacred while you're trying to pray to "get" some sacred. They don't understand that. They have no idea how to be there, and no idea of the reciprocity that is meant to exist between the people and that place. Because we are trying to protect this place, we find ourselves up there more often than even we should be, trying to protect it. We are trying to convince the Forest Service that you can't keep it a beautiful place if you let a stampede run through it.

Some people don't know how to appreciate a place without getting right down into it. The spring is very sensitive, but people don't think about this. Instead it's, "Yeah, I want those bubbles on me. I want to get right in there." They put crystals in it. They put Tiki dolls in it. Lately, people have been putting cremation remains in our sacred spring. My Grams said that you don't want to be buried in a sacred place. You can never rest if you're buried in a sacred place! Now those people will never rest. That's why we have burial grounds.

If we weren't going up there to collect all the crystals that people leave in the spring, it would be full of crystals and pennies and whatever else they put in there. Where do they learn this? I don't know where people learn ideas like, "We should leave this right here in the water." Do they think they are blessing the water? You can't bless the blesser. The spring is the sacred.

This all started when the Winnemem filed a lawsuit against a company that wanted to build a ski lodge and ski runs on Mt. Shasta that would have destroyed the spring. At the time, we exposed that the spring was a sacred pool. I think back to that as a mistake. We made a mistake saying it was a sacred place. Before that, people didn't know it was there.

Another challenge we had was that because it is such a sacred place, we never camped there. We don't have any artifacts there. We don't have any "evidence" because in our religion, you wouldn't leave anything in a place that sacred. That place is for the sacred beings. So, you wouldn't make arrowheads there, you wouldn't leave fire-cracked rocks there—or anything. It has been difficult for us to defend the right of our tribe to that sacred place without having physical evidence that we were actually there.

When we revealed that it is a sacred place, it turns out a lot of people were listening. Now, as a known sacred place on the mountain, it is hard to protect.

But, if people want to be a part of a sacred place, it's a big responsibility. It requires people to change a lot of things about their life, which they might not want to change. Does it make sense for a spiritual seeker to go to a beautiful place to try to be sacred for a minute? They want to have that privilege, but it's not a privilege. We Winnemem see ourselves as having a responsibility to and for that place.

HECHT: *In Toby's film,* Pilgrims and Tourists, *viewers witness the first time that Winnemem people ever saw the spring at Mount Shasta go completely dry. It is a terrible moment.*

SISK: Definitely.

HECHT: *I have heard that, since the filming, the spring has dried a second time. My understanding is that if it happens a third time, the spring will no longer be sacred. What does that mean for the Winnemem? What does that mean for the salmon? And the water? Can something become sacred again once it's no longer sacred? Or, by then, is it too late?*

SISK: I think that we're supposed to get the message now, and we're supposed to start undoing some things now. Different decisions need to be made at the Department of Water Resources. Industrial-scale farms need to stop using so much water. The fossil fuel mining and fracking companies have to stop changing all this water into waste. Water bottling companies need to stop taking so much—and they contaminate half of what they take. It is all for profit.

It's appalling that in the time between 1850 and today, all of the rivers in California have become polluted. If you go into the high mountains, they say to pack your water because you can't drink the water. That should be appalling to people! People should be up in arms about what we are doing to water. How do we clean it up? But, the response I see is more like, "Okay, we'll just filter it. We'll just bottle some water." It's a ho-hum kind of thing.

On the Klamath River, dogs get sick if they go in that water. The algae blooms are creating health issues. People can't even get in that water. Yet the attitude is, "Don't take those dams down. We need those dams." They don't associate standing water with high temperature water, slow running water beyond the dam, as creating a prime place for algae blooms. Yet those in power don't want to stop it because they have been trained to see standing water as the place to get water without understanding the other consequences.

Too many people want too much money. It's like there's no cap on capitalism. When do you get to that point that you don't have to worry about money anymore and you can just have a good life and do things as you please?

And now we find we are at a time that we can't afford. Can we afford another fracking mine? How many more water bottling companies can we afford? We know that they're going to contaminate half the water they take. The rest they are going to bottle and ship out. Who knows where it goes. Can we afford a drought state? Does anybody really know?

And when do we recognize the spiritual nature of water—something that we need to take care of and respect, not just use and sell?

HECHT: *This makes me think about your descendants. Through prayer, Grams is guiding you. You can go to the mountain to pray, and you can go to the round house to call on the Big Heads. What about your descendants? Can you feel their pull on you, feel them calling to you?*

SISK: It's all in prayer. The Big Heads come in from the stars. The stars above Mount Shasta, the Milky Way, send down these helpers that we call the Big Heads, that dance in the round house. When you pray about these things and a response comes, you then understand that these are things you also have to provide for future generations. Hopefully, many of the things we have spoken of will be in place before my generation is gone.

I imagine what Grams went through, being born on the river and living on the river. And then seeing the river flooded by Shasta Dam. She knew up and down the river—who lived there, what sacred places were there, what fishing grounds were there, what practices they did. She watched all of that go under water when the dam went up.

I was a future generation when a lot of that happened. And Grams led the Winnemem people in a way so that we had something left. A lot of tribes came out of this time with nothing. Even though the Winnemem didn't end up with any of our own land and we were subjected to the same treatment that all the California Indians were, we were never disconnected from the sacred places—those that were not flooded.

That keeps us closer to the land and our relations with all the plants, animals, trees, mountains, and sacred rocks. We still go to the spring, and we sing. There is nothing else but the prayer and the practice and the right and the responsibility to go to that spring and sing.

That's the job that we were given. And it's a hard thing for other people to understand because the language that we speak mimics the sounds that are found in this area. Our songs are the vibrations that tell the trees, the water, the Salmon, and the rocks that the Winnemem people are still here. And that we are still here to help each other coexist.

POEM

Promises, Promises
Frances H. Kakugawa

You promised me
A world, free of battlefields, soldiers, children
Abandoned in fear—hunger—
Children, trembling in closets, amidst gunshot fire.

You offered Hope, again and again.
A world, you said, where we will stand
Hand in hand, safe, beyond color, religion, gender, age,
Agenda—politics free.

You promised me a world
Free of poison in oceans, earth and air.
"You are the future. Come and be born in this world I will
Create for you. Trust me." You said.

My brothers and sisters who believed you
Are now old men and women, and they wait.
They wait.
They wait.

Stop using me, your invisible child
For meaningless promises and rhetoric.
Candles and flowers now fill spaces
Where my friends once lived and played.

Hear me.
See me.
The future is now. Today.
Today. Save me.

✴ 3 ✴

Healing

I am a colonizer and I am the colonized. I am a child of the excluded, this ritual unites us all.

MANEA SWEENEY

POEM

"To Future Kin"

Brian Calvert

To future kin? I guess I'd like to say
A thing or two before I'm dead and gone.
We all must love the world in our own way,

But children I have none, a choice I made,
Considering the damage we have done.
Still, had I future kin, to them I'd say:

Make the ridge as dawn defers to day,
Mark the eagles hunting in the sun,
Learn to love the world in your own way;

Seek beauty, feed birds, breathe, sit still, track game,
Leave wild things better for your having come,
For they are kin with many things to say;

Accept long winter nights, age, and decay,
Your time here will not be enough,
We all must leave the world in our own way;

Swim naked when you reach a mountain lake,
Dry naked, too, warmed by the dying sun.
To future kin, to one and all, I'd say,
We all must love the world in our own way.

ESSAY

Moving with the Rhythm of Life

Katherine Kassouf Cummings

It begins with a dance. Eight hundred men moving through the streets of Chicago. It is late August, the time when a rush of warm wind would send the switchgrass singing, when bees gather at the purple clover's plume, when martins dart along the lake's edge, scooping up dragonflies; a time of year when life erupts in this place.

The air is thick and still. Perspiration drips down the dancers' necks; steadfast, they sway through the city creating a river of human movement. Their feet caress the ground and their voices brush the sky. Together they sing. They shout. They dance for what they have lost and for what they are losing.

This was possibly the last time the Earth in this place felt the tread of humans who lived by life's rhythm, who understood what it meant to love the lands and waters, to fully love themselves. Those dancers were the guardians of the land where I live: Bodwewadmi, Pottawatomi, People of the Fire, on whose hunting grounds my first home stood. One hundred and fifty years before I arrived, the Pottawatomi people were driven from this land. The presence of absence lingers. What happens when you are taken from your home? How do you know where you belong?

The French word *dépaysement*, "the state of one who is out of place," expresses the disorientation of being away from one's homeland, acknowledging the connection between well-being and belonging to a particular place.[1] Each of us can trace our lives, through generations, to this original kinship. Though we may not know their names, we all have ancestors who knew the lands and waters as their relations, and we all have ancestors in the land. Dépaysement rests in our bodies. We long to belong.

As a girl, I wondered what it meant to be Lebanese American when I did not know the sound of the sea nor the scent of cedar trees; I wondered what it was to be Polish American, too, born three generations away from the land called Poland. My mother advised me: "Bloom where you are planted." America. And what is this "America" brushed with the names of the lands of those who walked before me? Some spectacular soil sprouting the seeds of countless cultures? Or a tough veneer resurfacing centuries of belonging? Assimilation is a complicated contract. What do we exchange when we leave the land that we know and to which we are known? The path of the dancers has been dug up and paved over. Does the rhythm beat out by the dancers' feet still throb within the earth, under the concrete?

Beyond the city of Chicago—a place now palpating with constant construction and countless cars—thrives a patch of prairie and a woodland where I spent many childhood days. There I discovered that life fluttered in the silky shimmer of a salamander. I felt the warm rush of awe under the wings of kestrels who sailed over golden grasses. My first friendships formed in this place. Though I would return to my bed each evening, to a neighborhood shaped by concrete, my heart pulsed to the prairie rhythm.

While I did not know it then, this prairie community flourished because of fire. Fire nourished, cleansed, and stimulated life in this place. The abundance and diversity of the prairie creates a bounty of decay by summer's end. A sudden blaze of accumulated thatch returns nutrients to the soil so that growth can begin again. This rhythm—fuel amassing, fire transforming, growth renewing—sustains the balance of the prairie.

The prairie that I played in as a girl exists as an apparition of what was; much has been lost here; much more lost before I arrived. The ground has gone cold, plowed under, paved over. And with the loss of the prairie, we've forgotten the friendship of fire. A violent amnesia, forced forgetting, draws us away from the rhythms of the land. The people who knew how to work with fire were removed from their communities at great cost to the balance of life in those places. Ecologist Stephen J. Pyne reflects on how this happened:

> [The Potawatomi] were known variously as the people of the place of the fire, or the keepers of the fire, because they maintained the great council fire around which the regional confederation of tribes gathered. But that fire did not stay within the council circle. It spread through the landscape, a constant among the diversity of grasses, trees, shrubs, ungulates, small mammals, birds and insects that congregated around the informing prairie. In time the Potawatomi became known equally as the people of the prairie since the one meant the other.

Remove fire and the prairie disappeared. Remove prairie and free-ranging fire lost its habitat. Remove the keepers of the fire and both prairie and fire vanish into overgrown scrub, weedy lots, or feral flame.[2]

Here in the place called Illinois, the prairie was largely taken over and turned under, but in the forests of the West, the living community continued its rhythm without the harmony of human participation. Shrubs and trees still grow and decay, but burns are banned. Fire appears an adversary to the ways we're living; it singes the semblance of permanence we crave. And yet, the ways we're living reel out of sync with the rhythm of life. As I write this, a shroud of smoke hangs over the mountains to the west. After years of fire suppression and fuel accumulation in much of the mountain forest, the frequency of fire and load of fuel are out of balance. Wildfires rage. Blazing beyond regenerative conditions for an ecosystem, these wildfires are a fearsome sight; homes destroyed, lives and land transformed or lost. Yet once, people participated in the cycles of bounty and burn. Those who walked before us, those who lived with the land and not merely on it, knew how to work with fire. They attended to life's rhythm; they observed the build-up of fuel; they knew when it was time to burn.

Today, we bear witness to a great unbalancing of life as record numbers of species are lost forever; the climate catapults toward chaos; glaciers melt; sea levels rise; the human population burgeons; people are pushed from their homes; and the crescendo of centuries of cultural dominance fills our ears. Does your heart race to imagine it all? How could we not—breathing beings, attuned to life's rhythm—feel the frenzied vibrations of so much unsteadiness? We have arrived here through the domination of worldviews that propagate disconnection. Losses that perpetuate greater loss unless . . .

Grieving serves as a cultural process for rekindling community well-being following loss. Inside a small, wooden church building, a community gathers to sing, share, dance—and grieve. Daughters have traveled here holding the death of mothers; mothers have traveled here bearing the death of children; families arrive carrying the loss of the places they called home; others mourn the life that has already left, the elders who never arrived, the extinctions that ripple onward, the ocean poisoned by plastic not far from where we sit. Today, this is my community. Today, we sit like schoolchildren on the first day of a new year, straining to start, knowing we cannot know what's ahead. As our guide reminds us, "Grief is a core competency that none of us were taught."[3]

Bowls of water and rock, old photographs and small things—a pair of gold earrings, a worn, wooden box—rest on scarves, bright blue, pink, yellow, orange. Roses and orchids keep company with sunflowers. I circle this bright shrine, conjured out of loss. I light small candles and perch them by pictures: a girl in a red sweater, smiling; a family snuggled together on a couch. This beautiful curation allows us to enter the ground of grief. The round, steady beat of a drum brings us into ritual, following old rules and ways. We thank each person when they finish mourning. The grieving each one does sustains the health of the whole community. By moving our personal grief onward, we lighten the load for everyone; by healing our hearts, we revitalize our relationships; we strengthen our community; we cultivate places for more love to grow. And so we raise our sorrow—together—we open our hearts, we allow the grief to flow.

The ritual ends. After grieving, my breathing becomes smoother, softer. I feel my heart lift on the wings of the lungs, which open with ease. As I step outside the sanctuary into daylight, misty rain washes away the dried tears crisscrossing my face. Raindrops tap my open hands with a rhythm reminding my human body that all this loving, all this grieving—like the prairie burning—is *how life moves*.

We are made for loss. As beings who love, grief is part of our life's work. The fibers of love and loss cannot be separated without painful consequences. When we lose what we love, we can become trapped under the thickness and weight of our grief. Our hearts stiffen in silence and stillness—until the fire comes. If the fire comes.

Grieving is not a given today. We've largely forgotten human rituals for rejuvenation. Once separated from the vibrance of community—from the homelands we were born to belong to, from the ways we once related together—fear, the great disconnector, easily takes hold. Life's rhythm fades behind the cacophony of a culture striving for comfort, constancy, and control. Thus separated, we are more inclined to sink into anxiety, addiction, distraction, depression. We must practice to become skilled in grieving so we can better handle the loss inherent in living. Community ritual reassures us of our embeddedness and reminds us of our responsibilities in the family of life.

Just as regular burns bring balance to life on the land, so rhythms of ritual that honor our losses support balance within human community. According to neuroscientist Bessel van der Kolk: "[The] brain is meant to be in synchrony with other brains. Interaction with other brains fundamentally shapes who we are. When we cry, we're supposed to get a response, and when we laugh, somebody is supposed to laugh with us. Those are the rhythms of life by which the brain develops."[4] Our most disturbing losses, our traumas, fundamentally affect how we relate and remember. Such losses can easily disrupt the relational rhythm that sustains community well-being. As van der Kolk's work demonstrates, movement has the potential to reattune the individual following the experience of trauma. Likewise, ritual offers a way to reconnect to community, to heal the heart in harmony after the great disconnection dealt by loss. And the process of grieving plots a path toward our ancestors, all the love that came before us. Perhaps it begins with a dance.

Our ancestors remind me that fire is not to be feared. Fuel amasses. Yet, no one plant accumulates all of the prairie's fuel, just as no one person shoulders all of life's loss. We have human family—relatives down the road and around the world—who will meet us in the fire. But we must attend to the fuel. We must attend to our grief. Each of us must have the courage to enter the flames. We must choose the burn. When what we loved erupts in ashes, what will grow?

ESSAY

(A Korowai) For When You Are Lost

Manea Sweeney

The flax harakeke heads were in bloom, blood red with wormy yellow-tipped rods tantalizing the passing birds to spread their seed. They were as fierce and rigid as the two people facing each other in the long grass. Their fighting had reached a crescendo, with rotten words flying across my head. Memories of old bones hurtling like stones between them, while I focused all of my attention on the flax flowers in front of me, crouched low enough to avoid hurtling objects and insults.

It seemed my parents liked fighting out here, in open fields just like our ancestors. My parents with sharp tongues not sharp *taiaha* wooden sticks to pierce their skins and carve the earth. The battle of this day would not be a story that would be retold in our songs and dances. It would be the start of hearing each other's truths, an impending divorce. Our long bus trip where joints stiffened with temperament, being dropped off here 20 km away from where we needed to be, the beating sun, and the heavy luggage, they were all only excuses for this fight. To scream away the fact that Nanny was gone. Dead of a heart attack. Another piece of our *korowai* cloak sewn into history.

In their fight, my father's forebears sat as prominent agitators around him in my mother's eyes. The colonizers, powerful men who took our land, stripped back its skin for production, and took its *mauri* life force, the oppressors who strapped her, her sisters and brothers in school for speaking our *Te Reo* language, and the men who enticed our families from ancestral homes to the cities, to drugs, to alcohol. The men that shot Indigenous peoples for sport while laying railroads in distant countries before returning "home." The psychological trauma inflicted by her own father, a man who had lost his way and his connection to where he had come from, and what it means to be a good human being. My father represented it all. She found a way to scream this truth of hers, as death so sharpens our focus. In that moment he also represented to her the Western doctor, too preoccupied with a loaded list of diabetics, and patients sick from their environs—too busy to notice that he had prescribed Nanny the wrong medication, pills that had given her heart a fright.

I edged under the shade of the flax bush, lay on my back, and looked up at the long dark green leaves as the screaming fell to sad murmurings between them. I remembered the times when Nanny and I had gathered flax at dawn, under thick fog before the sun crested the hills. She would *karakia* pray to the flax's parents and lineage, giving thanks for what we were about to take. She would make me recite the *whakapapa* genealogy of the baby leaf nested in the middle of the plant, the *rito*, and the *awhi-rito* the parents—she would tell me how they were a family, the *whanau*, and you can never break the *whanau* or the plant would die. She taught me to use the flax for sustenance; weaving bowls, knotted lines, and *korowai* cloaks for warmth. Every strand we wove represented a part of us. The flax itself had a lineage, meaning it was part of our ancestry, the interconnectedness of land, water, and us, an unquestionable relationship. I learnt to cut the flax at the base with an outward slant so rain water would run out of the plant and not drown it. We only ever took the long leaves around the outside. The extended family, or the grandparents—for they already knew their time had come to be part of someone else's story.

Tupuna ancestor I say aloud to the flax above me. Rolling the word around in my mouth, it tastes like wet earth. *Tupuna* I say to the rushing wind through a swelling throat. Those spirits who have passed into the underworld, through fire and flames to where gods tell them stories of those yet to walk our earth, and where time implodes on itself. *Tupuna* I say to my Mountains' dormant shadow in the distance, whose dance across this earth etched rivers and lakes that feed the ravenous *mana* and hearth of our land.

Tupu: to grow
Na: to originate from
Tu: I stand firm, in this place
Puna: a sacred water, new life

My back rests against the land, I reach my feet towards swaddled blue above, like *Tane* the Forest God when he pushed *Ranginui* Sky father and *Papatūānuku* Earth mother apart to first bring light into our world. Our *Rangiaatea*, the boundless unconsciousness nourishing the consciousness. My anchorage sweeps around me. Nanny is now a *tupuna*, she is the past and the omnipresent.

My father does not look at my mother again during that long walk to our *Marae* meeting house. His hurt eyes seem to be searching through his unanswered questions. Can he dispel this wrath inside of her? What must he do to mend so many wrongs that he did not do? Is he responsible for making this right? He had seen the pain that perpetuated itself and strangled her peoples, what authority but theirs alone could free them? He had seen her reading her great-grandfather's journals stolen from Nanny's cupboard, many years ago. Learning more of her story, finding gaping holes as well. It has been five years since she had stood on her ancestral lands, seen her mountain, felt the waters of her river running across her skin. We heard the shame in her voice, that the thread of this connection had been frayed, the day she got the phone call to say that Nanny had died.

We arrive at dusk where the dark faces in black clothes blend into a bleak night. Dragging our belongings behind us, my mother poured out of her all the insecurities in the form of instructions. Stand here, hold this, don't do that—as the formalities of tradition and ancestry overwhelm her. My father's kind hand finds the way to my shoulder, rests there in a calm silence, willing me to do the same. We greet people we do not know as *whanau* family. We join a group at the gate of carved wooden posts in front of two buildings. I see the river running past the *marae* and our *Pa Harakeke* flax bushes lining the stream. Ancient flax planted by ancient hands to sustain people who were yet to work our soils. Lush green leaves extend from their swampy bases, a bellbird sings from the seed stalk, the dark brown seed pods in full bloom to honor the one who has passed.

The old woman in front of the meeting house starts to wail and call words in a rhythmic chant, swaying. The woman at the front of our group calls back, slowly we walk forward. This happens again, backwards and forwards, backwards and forwards, one chanting strand over the other weaving us together with words. They speak of our genealogy and why we are here. Acknowledging the dead, honoring the living, tracing our heritage from mountains to sea. A sound comes from a deep hollow part of my mother's stomach. Clouds have crept in to ease the heat, filled with fat raindrops from the day's sweat. They fall, each pounding the fern root that covers our valley floor. The sound of this echoes off the hills. This is the name that was given to our land by our ancestors—the sound of rain pounding fern root.

In the dim light, against the back wall of the intricately carved interior, I see a black casket with an open lid. I see Nanny's flax cloak draped over the end of the casket adorned with dust colored feathers. The tears stir inside of me. *Ko wai au?* Who am I? My mother weeps, she weeps because it feels good to be back home. I look at my father sitting in the back row concentrating as the *whaikorero* speeches begin to honor the dead. I am a colonizer and I am the colonized. I am a child of the excluded, this ritual unites us all. The dead are in no rush, and the living cannot bear to let go—and so the speeches and stories of Nanny continue late into the night.

A *kuia* old lady grabs my mother's hand. "E kui aren't you Neke's daughter?" She whispers. My mother nods, leaning closer to hear the woman. "Aue e child, look at you. So beautiful like your mother. See koro over there? Let's have a cuppa tea with him later on, he'd love to hear what you and your whanau have been up to." Mother nods again and sits back on her haunches, exhaling the tension of one abandoned. That evening, we talk to aunties, uncles, cousins, those unfamiliar who greet us with their story thread that connects us to each other. Despite the long absence, my mother's *ahi-kaa* still burned within her, a fire that needed rekindling. Never did our presence get questioned. We are all indigenous to somewhere, something.

That night a fitful sleep took hold of me under framed faces of *tupuna*, next to Nanny's lifeless body. A dusty road alone, on another swelling, record-hot summer's day. Old run-down houses, their paint pallet of pastels, faded like washed-out watercolor paintings. Old cars litter driveways bleeding rust, twisted metal, a baby alone in a front yard, the smell of marijuana, towns dusted in gravel grey. Curtains of rain pulled across bare earth. A sea swallowing *Pa Harakeke*. Filthy water and floating fish. An abundance of hunger, an absence of food. Fighting against our own histories of ties we cannot sever. Silence in spaces where bird songs existed. Pained shadows of *tupuna* past. This is not our future I say. This is not our future I scream.

I drag the *korowai* flax cloak off Nanny's casket and wrap myself, turning to face my sleeping mother, eyes drifting inwards. A great wealth fills my heart. I travel to where queens, kings, and dynasties ruled, to stand in land that is also mine. I witness the tentacles of technology helping us to understand old stories in new ways. I return here to see fields of flax surrounding healthy villages. The healing of multitudes has begun in a climate that is rebalancing from the torrential downpours and fierce winds. The land is tended as *tupuna*, gratitude and moderation is practiced as a tradition of the resilient ones. Connections are shared, storytelling is complete and not withheld from those who need to hear it. The resilient ones do not need to know their stories of the past to complete them. They are already their fabric, textures and hues to marvel, with a confidence in their validity, knowing it does not rest on their consciousness alone. These stories are already becoming, celebrated in the multiple strands that weave together their *korowai* cloak.

We return Nanny's bones to earth, as the sun crests the hills in the morning fog. Her memory linked with ancestors of distant lands who had the courage to practice old traditions in new worlds. People leave, washing their hands with cool water to rinse the sacredness of this ritual from them and return them to the land of the living. I remain and linger in this in-between place, feeling the *harakeke* seeds rolling around in my pocket, hopeful of their new future. Squatting, I push the seeds into the damp earth. "E kore koe e ngaro e kui, he kaakano i ruia mai i Rangiaatea. You shall never be lost, we are all seeds scattered from Rangiaatea." It is a hopeful farewell, with the singing of the resilient ones calling me into a powerful future.

ESSAY

To Hope of Becoming Ancestors

*Princess Daazhraii Johnson and
Julianne Lutz Warren*

WARREN: What kind of ancestor do you want to be?
  JOHNSON: Let me start by saying . . .
I hope that I am an Ancestor to someone.
  And, in the most blessed scenario that I am an Ancestor,
    I want to be the kind of Ancestor whose song story carries on the wind.
      WARREN: Are songs free for the taking?
    JOHNSON: I want to be the Ancestor who crackles and spits sparks
    from the
fire as a reminder to stay alert and follow dreams.
    WARREN: What makes for a worthy dream?
  JOHNSON: I want to be the Ancestor gently guiding the lost and healing the
  wounded.
    WARREN: To where? And, what is your medicine, the source?
  JOHNSON: I want to be the kind of Ancestor whose laugher echoes through the
canyons and lands as a snowflake on an eyelash—a
  beautiful and fluid humor.
    WARREN: What makes you laugh?
  JOHNSON: The many questions you ask are for the reader to decipher.
    What kind of Ancestor
    do you want to be?
  WARREN: First, if I may, your vision just raised more questions to mind. I
know that you
  and your husband have three beautiful kids. So, when you say that you
hope to be an Ancestor,

are you saying that this function doesn't necessarily come with parenting?

I have no children, which has felt like a vacancy. But, I haven't thought so much about

the reverse, that having babies might not automatically make you an Ancestor.

Maybe you're saying there is a difference between being an a̲ncestor and an Ancestor?

JOHNSON: I don't believe that one must have biological children to be an Ancestor.

Not at all.

WARREN: Can Ancestors be adopted?

In the scenarios you painted of the

kind of Ancestor you hope to be,

I noticed a common spirit:

Whether as **singer energizer healer laugher**—you are a giver—of **stories dreams belonging good humor.**

You don't give alone. Each of your blessings depends upon an intimate partnership

with other elements of nature—

of **wind fire ground water**

to **carry spark orient stream** your gifts.

What you say also reminds me of a film image.

The film is *Drums of Winter* produced by Sarah Elder and Len Kamerling in 1989. Maybe you know it?

JOHNSON: Len was my first screenwriting teacher. And, I am familiar with this documentary though it's been so many years since I last watched it. Many of the films I have grown to love are because Len introduced me to them.

WARREN: In the film, there is a Yup'ik woman named Lala Charles. She grew p there in Emmonak, Alaska.

She is describing her first dance and imagining how a parent might feel looking on at her child. She says:

"The parents ... I wouldn't know how they feel. ... I guess if I was them, if my child was dancing, I'd get up there and dance with them ... I remember mom ... was dancing right there behind me. I could feel that she was dancing as hard as she could."

WARREN: The frame shifts now to her past—a young Lala enters shyly into the center of a circle of people. Her parents stand behind her, not drawing attention to themselves. It is as if through the very palms of their hands or their eyes or breath they are helping her do her very own best.

If I get to be an Ancestor, this is one kind I'd want to be.

JOHNSON: That's a beautiful way to imagine being
an Ancestor. Guiding, dancing with, and encouraging
the young child.

The screenwriter Waldo Salt likened the relatively new form of
screenwriting to poetry. Poetry conveys the image and feeling.
Show don't tell.

Not everyone likes poetry and I
often hear because they don't get
it. But the beauty of poetry for me
personally, is the journey we take—
and in the case of really good poetry—

the deliverance!

The most important thing to concentrate
on is your own soul. Wise words from
my late grandfather.

I believe my Ancestors are helping guide my soul
(currently) through this human life that is only one
segment of a journey,

a journey whose origin I do not know
or can't remember
and whose forward projection I have
yet to discover.

WARREN: There is a particular intimacy, I feel, when you recognize that a person closely related to you biologically—that is, someone with whom you may share a name and even the same nose—is also an Ancestor in a chosen sense.

Sometimes I hear a line of my great, grand uncle, John Burroughs, repeat in myself. In a quieting way, his poetry speaks,

"What is mine shall know my face."

WARREN: While this uncle seemed more openly tender than many men of my culture, yet, he also enacted all-too-typical, life-shattering ways toward women. Too many of my ancestors' stories disquiet and silence others.

There is, for example, also this to-me familiar saying, "children should be seen, but not heard."

When I was a kid, I very much wanted to be heard. And, seen.

I've grown up in a colonizing culture colonizing me into submission—kid to adult, woman to man—and into domination—church over non-churched, white over brown and black skins and over others' lands. Consumption. Of. Lands-and-Peoples. Money in exchange for water and earth.

I am afraid of passing on these attitudes.

                      I want to unlearn them.

I don't intend to fail, but where I do, I would want rising generations to **refuse** doing likewise, even, if necessary,

to **refuse** knowing me as an Ancestor. Robert Pogue Harrison's book *The Dominion of the Dead* has helped me think about some of this. . . .

To be the kind of Ancestor I want to be, then, I must learn how to show all possible respect while talking back to some ancestors who are would-be Ancestors.

I must learn to resist—in part or in whole—their heritage.

                    JOHNSON: What a thought—an Ancestor that
                    should be resisted if/ where her
                    legacy is harmful.
                    Never considered that.

WARREN: . . . This leaves lacunae . . .

JOHNSON: We are living during a time of broken   silences. The feelings and thoughts of women, of Indigenous people, of Nature, of all who have been "othered" are

    breaching the colonizers'

dams of power. We must remind ourselves that we are here for a reason and as long as

        we continue to breathe this sacred breath

then we must do all we can to restore balance to our lives and the world around us.

        WARREN: The sacred world around us, also breathing *for* us, if I listen
           I hear, into the cracks   and   ruptures . . .

Right now, out my window, I see
a fox who has become familiar.
Sometimes she appears this time
of day. She's tilting her ears,
a little sniff. Ha, she squatted to pee.

She lopes back into the woods.
> I feel her company gratefully. I get this longing. It could be loneliness. Maybe she is teaching me to become more receptive to many other Ancestors I have been missing—not only human ones—who, if I paid better attention, are revealing themselves, passing along gifts with potential to help me become a worthy Ancestor.

(For now, a side note: I am learning more of my maternal lineage, the spirited Frisian folk of the North Sea coast, now the Netherlands . . . They've fought hard for the language, which means something to me . . .)

JOHNSON: When I say I hope that I get to be an Ancestor to someone . . . I am considering the future and the ability for our Mother Earth to host human life.

WARREN: O! I went along on some twisty trails from your first statement, didn't I?!

Speaking of laughter!

And, speaking of being the sort who "sparks . . . a reminder to stay alert and follow dreams" while "gently guiding the lost and healing the wounded . . ."

You are bringing up whether the fecund health of "Mother Earth" is resilient enough to withstand her abusers.

JOHNSON: Still,
> the Yukon River and the mountains surrounding Arctic Village are also Ancestors. Inhabited by an ancient energy. The thought they will always be there brings
>> incredible comfort.

I've always thought that technology advanced without a spiritual consideration to what it would all bring.

While there are so many benefits to technology,

it has also taken us away from the land and animals—away from keeping our hands working with the soil and the elements around us. It keeps us in the most unnatural state—glued to screens for hours upon hours.

Hours that I imagine I could be learning how to make snowshoes or birch bark baskets or being out on the land
with my children.

(This is reminding me that I need a more ergonomical office chair! Haha!)

WARREN: (Haha! Writing can be back-breaking work; I think that is underappreciated!)
For sure, human powers are complicated.
    I keep returning to this word *hope*. Like any technology, *hope* unmoored from Spirit or divided from the Sacred, can take more than it gives, I think.
    Hope is a tricky one.
    I've been contemplating two Old English meanings of *hope*—abstract and grounded—and how they also might merge into one.
    In the abstract, *hope* is some mingling of expectation and desire.
        Expectation   implies some level of confidence in someone or something. Confidence, however, can be erroneous and desires can be more or less worthy.
So, not all hopes are of equal merit, in my opinion. Whereas,
    in the grounded sense (though lately fallen out of use), *hope* designates an actual place that is a haven or a refuge. For example, an island in the sea, a soil-
rich valley surrounded by craggy mountain (whose nooks might nest eagles' eggs safely), a
womb nourishing a child, and, for a child learning to dance, inside a circle of loving hands.
    Earth, in this way,
        is a blue-green Hope-of-hopes surrounded by uninhabitable space
full of stars out of whose dust the whole, alive planet is made.
    Idealized and real hopes, then, would seem to merge into conditions of
    health when
human beings bend with "ancient energy," as you put it so powerfully, "inhabiting"—rivers
and mountains (come to think of it, also indwelling all ephemeral earthlings, right?)—a proven re/generative force warranting both trust and desire.
    While, trying to control the Ancestors, or, to somehow own the Essential is a madness of pride and selfish desire—that is, a hope of terror—that   cuts us off,
        also from future thriving.
    Any hope to become an Ancestor ("to go before") is lost
if the children are lost.
    Here's a
downer: A womb, heartbreakingly,
can become a dangerous place, or even a tomb.
    There is a mother, perhaps she has no other choice—

there are many mothers—none of whom can escape
> eating poisoned food webs
> drinking toxic waters
>> moving within a less-and-less habitable climate of "Mother Earth"—

Womb-of-wombs—
> with abortifacient consequences, although no
> surgeon's   knife is in sight.

    JOHNSON: I heard a young Sikh woman, Valarie Kaur, recite a tremendous poem—a take on Trump's America and she made that analogy
> not from womb to tomb
> but from tomb to womb.

That we cannot give birth to the new without
the pain of
> TRANSITION.

                WARREN: WOW. Speaking of   deliverance   I feel it.

    JOHNSON: I hope in this short time here, I am able to express my love for our Mother Earth
> not just in word, but deed. And that that expression lives on in the

generations to come.

        WARREN: Me too.

"Concentrate . . . on your own soul," your grandfather said. I am hearing
the movement of your opening vision—
> **singing   spitting   guiding   echoing...**
>> Is this movement skillful love?

    It also occurs to me that we've been talking about movement, in many senses,
        all along.
There is your belief that your human life is a segment of a journey. And, that is not
unlike a
poem, maybe. Lala growing up as she dances. My call
>> to listen forward

even while I talk back.

    Also, our conversation paused for a span   while we both traveled,
      as it so happened, each of us to beautiful islands.

                JOHNSON: It was a blessing and privilege to spend our
>> holidays in Hawaii.

            WARREN: That sounds wonderful.

    JOHNSON: Time away was
> necessary—
>> to see my

                              children playing in the
                              ocean. To feel the sun
                              and soak in the laughter
                              of friends and family.
              WARREN: I wonder, can time, like place, be *hope*?
    JOHNSON: I have no way of answering that. I try not to think of time in terms
of past and future.
       What brings me the most peace is simply being present in this moment.
                     WARREN: Aaah. Thanks for that. Yes . . .
Wow.
       Now I am imagining this whole Earth, as a place, yes, but also as
       one moment,   as in, hey, maybe,   a *Spell* . . .
       And, maybe this Spell endures from mysterious origin to undiscovered
              forward projection, as you conceived earlier for your human life—
       an episode of a vaster story
       where death and life aren't so easy to differentiate.
                     To remember—to echo, or, to recompose—isn't that revival?
       Remembering is interwoven with my time away in Aotearoa New
Zealand.
       In his poem "Learning a Dead Language," W.S. Merwin says,
". . . To remember
Is not to rehearse but to hear what never
Has fallen silent . . ."

       My journey delivered me into absences of many birdkinds who were made
       extinct by
complex human causes—across time and space.   I was delivered into voids
       of feathers and songs with generations of manawhenua
              interrelations—including,
recently,   without huia—as the native bush was
              by new colonizers, in te reo Māori—*Pākehās*.
       I am more easily startled in new situations. Facing my own vulnerability
       helps me learn:
One thing I noticed was how similar a void can feel from one instance to another,
even half-
              a-world away—and, how hard it is and how much I wanted to hear into the
       Spell   "what never
              Has fallen silent"   still speaking in distinct languages
                                   outside my own head.

This listening, again, in Merwin's words, is to "cultivate the awareness" of what belongs
to no one, yet has been saved for everyone, and passed on, by others who have shared in it. Ah!
I suddenly recognize this as another way to say a kind of Ancestor I'd like to become.
 I acknowledge that I am making my home now within unceded Ancestral Lands of
Tanana Dene People, extending north and east into those of your People,
      the Gwich'in Nation—to
 the coastal plain that I've heard you call, in your language,
 Iizhik Gwats'an Gwandaii Goodlit (The Sacred Place Where Life Begins).
 Maybe there's a whisper of respect-breathing hope in the US claim on this ground, in
federal terms, as belonging to "The Arctic Refuge."
 Foremost, though, this designation is an ironic self-confession of a hope that menaces—
one that has not yet been talked back to enough nor transformed. Case-in-point, machines seeking to steal ancient energy, not to dwell with it,
wait at the brink.

 JOHNSON: I refuse to allow myself to think of these "machines" in the birthing grounds of the *Vadzaih*—the Porcupine Caribou Herd
 for we as Gwich'in are spiritually tied to the vadzaih.
 The human rights of all people are under threat, but it is those of us with brown skin at
the frontlines of the atrocities we are facing across the world.
 For all the peace in my heart, there is a also a fierce rage ready to Defend the Sacred.
 Together we are as powerful as thousands of thundering hooves shaking the earth.
  Our intentions—antlers
   Our intuition guiding us on an ancient migration
We can never forget
  WARREN: "We can never forget" I hear this echo
   Revival
binding shared Hope as a wound in your words, "a song story":
 Bar-tailed godwits are among the hundreds of birdkinds who nest alongside, for Gwich'in, *Vadzaih* (when I try saying your word it sounds like a wind).
 Come fall, these sandpipers migrate all the way round to the summertime of Earth's

other pole!

  Another writer, a younger friend, Manea Sweeney, told me the birds' name is *kuaka* in
te reo. And, she guided me to find them in Foxton Beach, at the
    mouth of the Manawatu River just downstream from former huia haunts.
(In some regions, I have learned, a place where kuaka gather is called
      tahuna a tapu, or, a sacred sandbank.)

  Joy overcame me when I heard a flock take wing chirping chattering. I felt their companionship drawn from where I had come from and would also return. Bits of
   their songs nested in my memory even while
kuaka language continued to speak on, whole and awake-in-the-Earth.

  Manea bent over and picked up a small brown feather that had
fallen on the wet sand.

    She stood up and handed it to me . . .

      JOHNSON: Where are we now?

What next?

      WARREN: Yes, where are we now?

        What next?

    As you said earlier, perhaps that is "for the reader to decipher."
Mahsi' choo, Friend.

      JOHNSON: Blessings along your journey.

INTERVIEW

Camille T. Dungy and Crystal Williams

DUNGY: I regularly wear a ring that belonged to a great aunt (a social worker in Brooklyn) and often wear a coat that belonged to my grandmother (a librarian in Chicago). Both fit me nearly perfectly and I love knowing that the body I am in resembles the bodies of women who came before me. Women who used their lives to support and lift up the lives of others. Looking at these wearable heirlooms, I am reminded how little things can make big differences in many people's lives. Aunt Birda's legacy is with me daily, not just because of the ring but also because of a drive I share with her to focus my life towards the encouragement of others. It's like I carry on her legacy not just by wearing her ring but by living my life with an attention toward justice and a greater social good. What about you, Crystal? Do you have any tangible objects that you might use as a touchstone as you think about who you want to be in the world?

WILLIAMS: I always marvel at the immediate family histories you carry with you because as an adopted person, other than my body and the histories it carries, I haven't any heirlooms from my biological family. Of my parents—the people who raised me who have both since died—I have kept little, although I still have my father's Masonic ring and my mother's wedding ring. They are both in a box that I almost never open. What I hold sacred are a series of photographs of me as a baby. In them I am exploding with joy. They are on the wall in my bedroom among a series of photos of my parents and other family members. When I look at the baby photos, in particular, I am reminded of the kind of faith in the universe my biological family must have had in order to give me up and also the kind of faith and love my parents had to take in and tend to me in a way that resulted in such chronic joyfulness. I am reminded that we—people of African descent, despite what the world tells us—deserve joyfulness. And I think of both of these kinds of love—faith and tending to—as legacies. These ideals are my touchstones. They are my guides for honoring not only the biological families to which I am physically connected but do not know and my parents who no longer walk with us, but also our African ancestors whose faith and tending allows you and I to be poets, humanists, writing together here and now.

DUNGY: "Chronic joyfulness." I love that phrase. It is so important to think about the ways we can live our lives in a perpetual condition of joy. But this is not a simple proposition. The world is full of horror, sadness, sickness, devastation, and malice. Much of that malice is directed at us—as women, as people of African descent, as artists, as educators, as . . . as . . . as. . . . Still more of the malice glances off us, as if we and those for whom we have come to care are merely collateral damage. As a person who has chosen to live a life of attention, I can't help but to see the chronic pain all around me. Often, I suffer as a result of my awareness. And yet, I insist on joy. I insist on discovering ways to feed myself and, through that sustenance, to feed others. I write often of my garden, a ruckus of wildflowers all my neighbors—human, bird, bunny, and bug—are welcome to enjoy. I came to a decision recently to vocalize my appreciation of other people when such appreciations cross my mind. When I see a person with a pair of socks I admire, I compliment the person on said socks. When I think a streak of purple in someone's hair is great, I let her know. If a toddler in my immediate vicinity is particularly adorable, I tell the supervising adult. If a man's cane is pleasantly vibrant, I tell him. This demands a kind of opening up to strangers, a kind of momentary release of composure that allows me to express appreciation for lots of little things. Though it is occasionally awkward—mostly because people don't expect these kinds of interactions and are therefore sometimes clumsy with their responses—for the most part, my small statements of praise seem to result in three to five seconds of shared joy. Nothing we do in this world is easy. If I'm going to put effort into something, I choose to spread a little more joy.

I just described two intentional practices that keep me attuned to potential and to joy. Writing is another path towards this kind of "faith and tending." So are, for me, teaching and mothering. When I think about the elders and ancestors whose spirits and intentions guide me, I believe they, too, chose to recognize positive potential in themselves and in others.

That's what they mean when they talk about spreading the light. When Audre Lorde said, "Caring for myself is not self-indulgence, it is self-preservation, and that is an act of political warfare," she was thinking, among other things, about the revolutionary potential of finding ways to keep herself attuned to her own internal—and eternal—peace and joy. It makes me feel good when I spread light, and so I see the sharing of joy as a simultaneously self-sustaining and revolutionary act.

I always think of you, Crystal, as a person who walks in the kind of light I am trying to describe. I am grateful to you for this. What are some of your intentions in this realm? Who/what are some of your models?

WILLIAMS: As I was reading your response, Camille, I thought: I know she's not from Detroit, but maybe her people are . . . ? In Detroit, where I did most of my growing up, this "vocalization of appreciation" as you call it is a habit Black people regularly practice. It can manifest anywhere and from anyone in small moments, something like love-taps: "hey, that skirt is cute," or "your hair is fly," or "I appreciate you." It's a lovely practice I try to undertake regularly. I think of it as spreading a little love dust. And while that may sound silly to some, I agree with you and think of it as utterly serious because, you're right, not only does it spread something outward, but it reaffirms something within me, confirms my moral and spiritual center, reminds me of our interconnectedness and of our capacity to love everything, no matter how small and seemingly meaningless. So that's one intention. I love that we share it!

It delights me to hear that you see me walking in the mode of "spreading light." It means that lessons I've learned in these many years have taken and that I am accomplishing one of my central goals: to lift others up, in love. I want to help people be their fullest selves. I think that's when we're our most beautiful. I have had so many mentors who have done this for me, who have loved me so deeply, have been honest guides. What a gift! So, for me, this sort of lifting is an obligation and I consider the ability to do so a gift from our ancestors. If we do not lift each other up, who will? You list Audre as a guide for you. For me in addition to my mentors, the poets Lucille Clifton and Jack Gilbert were guides, and also Pema Chödrön. The gesture of her work rooted deeply in me nearly twenty years ago and continues to bloom. All three of these spirits suggest to me a way to exist in love, with rigorous self-reflection. Finally, I carry with me an old African spirit who appeared to me in a dream a decade ago and who remains with me now. She is gentle and loving and never speaks, but always communicates. And she often points me to the past and the people, the long line of people who stand behind her. I'm struck by your habit of spreading joy, Camille—your garden, your writing, your poetry, mothering. It's just occurred to me that I have always seen you as a particularly joyful vessel. And when I see pictures of your daughter, she too, seems particularly joyful. Do you think joy is an inheritance? How do you talk to your daughter about it, our ancestors? Did your mother with you? Mine did not. This has been a solitary journey towards and with my ancestors. And it causes me to wonder, more broadly, what the long-term impact of people not speaking to their children about heritage and gifts is now and might be.

DUNGY: I am honored that you think I might have some Detroit in me. What's probably more likely is that Detroit has in it the retention of all the places people brought with them when they came to Detroit. There are so many amazing retentions of culture and bearing that we don't even always know to be direct retentions. When I move through the world, what are the ways I am echoing the ways people in the past have moved, walked, talked, eaten, danced, sung, mourned, praised? That makes me think of what sort of legacy I would want to leave to someone who unwittingly followed my lead. It's one thing to think about what we purposefully pass forward. It's a whole different thing to think about what we pass forward without intending to do so. When they say, "Be the change you want to see in the world," this means be the change in *everything* you do—when you're putting your mind to things, and also when you are not. If you're not consistent, your thoughtless actions may erase the positive change you work so hard at creating with your conscious actions. So, that behavior of praise you recognize as something you cherish from your home in Detroit, I would imagine that this is something many Black folks can point to within their communities. It makes sense that this kind of active, joyful communion would be necessary to our survival and our flourishing and that you can find evidence of such behavior in places all across the Diaspora. We have repeated this behavior without necessarily realizing we're participating in a custom that was established generations ago. Joy, then, this love dust as you call it, is a kind of inheritance. I like to think that.

I do think that my mother and her parents taught me a great deal about cultivating the kind of joy we're talking about here. My grandmother was something like a walking ray of light. Truly, she brought joy everywhere she moved in the world. My daughter is named for her, and I often share stories so that she knows where she comes from. We have an ancestor wall at home, with photographs of people reaching as far back as my grandmother's grandmother (also my daughter's namesake). Frequently, we take some time to look at the photographs and say hello to everyone. I think that knowing these stories, knowing where she came from in this way, is important. But there is so much none of us can know, and it is important to acknowledge that mystery as well. It is important to open a space for the ancestors for whom photographs were not an option, for whom naming rites were not an option. And, also, it's important to leave space for those with whom blood connections might not be relevant. My daughter and I talk all the time about my mother's best friend, a woman who was another grandmother to my daughter and an aunt to me. She died last year, and we all miss her dearly. Having connections with the past, retentions from the past, doesn't require that someone be linked into your immediate family tree. It seems important to me that we open up to the possibility of connection with those whose strict rules might say we have no reason to call our own. I mean, I'm not from Detroit and I assume you and I are not related in that kind of way, but considering the Diasporic past, there is no real telling who is and isn't blood kin. More importantly, if it is true that matter is neither created nor destroyed, whatever matter makes me is, and has always already been, connected to you.

I'm trying to think whether this would have been the sort of thing my mother would have talked about with me. I don't know. I know it's something I've always believed. I was never pushed away from believing it, so that's important in its own way. I know it's the kind of thing I talk about with my daughter. I know that my family has modeled a kind of openness that aligns with this way of thinking about the world. Most of the pictures of the ancestors on my walls come from my parents, so that connection to the past is certainly inherited and cultivated by my family. But I think it's also true that I organize these ideas in a way that's different from how my parents or grandparents would. It's an interesting question: How much does what we've been actively taught about how to navigate the world with and through the ancestors matter for how we end up actually walking through the world and into the realm of ancestors ourselves? Do you think you'd do things differently if you didn't feel so solitary in this journey? Or do you think that the very fact that you have made this journey in a solitary manner has shaped your drive to be a mentor and guide to others in the many ways you do that in your life?

WILLIAMS: Your last two questions caused me to pause for a great deal of time and, I fear, I haven't an answer that I trust for either of them. Essentially, I hear them to be asking: Would your fundamental self be different were you not so solitary? And I just don't know. I think we are both nature and nurture and the way we live our lives is a kind of nurture. So it's possible I would be a different kind of mentor or guide, as you say, than I am now. On the other hand, it's also possible that this bearing is an inheritance over which I have little-to-no control, a legacy and imperative passed down genetically or spiritually or both. In that case, I'd be the same, only traveling with beloveds very closely beside me. I suspect it may be some combination: my natural inclination has been intensified as has my sense of connection to all peoples precisely because I have to-date been so solitary.

I find your memory of your grandmother so moving and of your mother's best friend. Both cause me to think that how we retain and revere the past has to do with how the people we love and admire hold or reflect the past in their daily lives, how they reconcile their own histories, how they model existing in love or in pain, in wholeness or as broken. And ultimately, I do believe that this is how our ancestors speak most loudly—if we care to listen. It seems you and I have been extraordinarily lucky to have had ancestors whose angles of vision on their lives were ones of magnitude and extreme compassion and love (which can include fight, just to be clear) and whose voices are clear and resound. I am so happy for your daughter and for hers. And in truth, I'm so happy and thankful for myself. How these folks buoy me! What are you happiest for in regards to the ancestors?

DUNGY: One of the things I am happiest for in regards to the ancestors is purely living with an awareness of their existence. I am happy to have shaped my life in such a way as to know that I do not walk alone. This does not necessarily mean something like what we understand with that footprints in the sand poster, where during the time when you only see one set of footprints rather than two it is because that is when God is said to have been carrying a person. I'm not saying that the ancestors' job is to carry me as some sort of permanent spiritual helpmate. That's asking a lot of people who have already done a lot of hard, and often painful, work. More to the point, I believe that the work we do now can be reflected in the world in all kinds of ways—matter is neither created nor destroyed—and I want to do the best I can to set up a legacy that will reflect positively on the life I lived. I'm not perfect. None of us can be, nor should that be our goal. But I'll keep trying my hardest to leave things better than I found them. I've learned that way of life from many who have come before me, as well as from people like you who walk alongside me in the here and now.

POEM

Yes, I Will

Frances H. Kakugawa

When you take my hand in yours,
Your tiny little fingers curled around mine,
I am filled with a great sense of duty,
Duty to keep this world
Free from fear and evil.

When I feel your hand in mine,
The contrast: spring to autumn,
I feel compelled to live
Every minute of my life
With love and human kindness.

So this world I leave for you
Will be safe and peaceful
A world, of freedom
To think, create, speak and define.
A world of possibilities—

With white cumulus clouds—
Cerulean skies, crayoned life in coral reefs,
Frogs and forests awaiting explorations.
Oh, I will live to make a difference
For having felt your hand in mine.

* 4 *

Interwoven

The land is a barometer for how we are living as ancestors, whether or not we are remembered.

JOHN HAUSDOERFFER

POEM

Alive in This Century
Kristi Leora Gansworth

To my child, beyond welcome, I might share thoughts like:
when you become aware, there is something
that happens. Maybe there are those
who have known all along, who have
protected fields of water and seamless streams
of medicines, the berries
and leaves and meats of unaffected relatives
who give all of themselves
for every ailment, every sorrow, every hunger. It is true that
you carry with you, within you, breath—
and in each of the earth's directions
you will encounter those
who know the worth
and meaning of this gift. And how beyond words
all of this experience is. It doesn't come so easy
to all, some lessons are accepted
only through pain. Maybe, somewhere,
I knew the dangers of this place, the gnarled walk
of bone-on-bone, studded skin
of scarring, gnashed cartographies
where we outlive terror
all these generations. Maybe I knew
we would be seen

from beyond, even loved
from the moon:
your grandmother, and mine,
we would be cared for
by earth's bones:
your grandfathers, and mine,
those who stood to form
our Anishinaabe miikana. We go to
that life, we come from that life. On escalators,
    in staircases, on trains and subways
        water ascends my spine and spills
            from my head. They come close when the calling
                is stirred, they come close
            from where they are, trails of ancient memories
buried in this land, buried in your body,
and mine. There is safety, there is comfort
when they breathe
into every rogue cell
that recalls: we are part of this vision
even amid the pain
of wrenched bodies, burdened souls,
times when
this life seems
an unending night
of dark, dewy flowers lit by
stark, glaring pulses of star
and cold, cold uncertainty. They call this
light, light of this planet, light beyond
the peripheral sliver of forever
we know as a lifetime, light
a form of travel
just as water is both liquid
and traveled path.

As certain as death
marks dissolvement, some
essence remains in mist,
spaces between what is visible and

what is known
in some other way.

Breathe, live—

When thinking of what life might tell you,
I am content with the simplicity
of these mysteries.

ESSAY

What Is Your Rice?

John Hausdoerffer

When I first studied climate disruption in the early 1990s, the data led my imagination to grieve for my unborn grandchildren, and for their children. Projected maps of ice cap recession, species extinction, sea-level rises, and human displacement stretched my moral compass to fear for the peoples, beings, and systems of 2050 and beyond. I never fully imagined myself living with these consequences, in my prime. The grave dangers seemed "away" in space and time, just like my food came from "away" and like my waste went "away." But as a father in 2018, all of that changed.

Every month of every year, my daughters—Atalaya, 10, and Sol, 7—and I like to go "snow hunting" in the mountains around our home of Gunnison, Colorado. Even in summer and early fall we strap skis to backpacks and scramble up slopes until we find those tucked-away wind deposits that cling to the mountain through summer's heat. Then we ski, laughing at ourselves for loving that dance down snow so much that we are willing to hike four hours for fifteen turns. Our rule is that we have to ski once per month, every month, until we get to one hundred straight months in 2023.

Snow hunting is a challenge that has also taught my daughters about their mountains, about those first trickles of water that lead from the snow patches of the Elk Mountains to the tributaries of the East and Gunnison Rivers until the Colorado River returns those waters to the Pacific Ocean and, eventually, until the cycles of life return the moisture right back home as new snow to renew our dance and the dance of life. Snow renews the cultures and ecosystems of the West, and it also renews my bond with my daughters. Snow is sacred.

Whenever we worry that we might miss a month or fall short in finding snow somewhere new, we look to Yule Pass, old reliable. Just west of Crested Butte, Colorado, Yule Pass sits at 11,300 feet, tucked between the dark slate of Purple Mountain and the orange scree of Cinnamon Mountain. Since 1992, I have skied the heart-shaped snow of Yule Pass in September. Atalaya has skied it five times with me, and in 2018 we were eager to initiate Sol. Sol was even more eager for the initiation, after having heard tales of Yule Pass for years.

But when we arrived, the snow was gone. There was a small, dirty strip on the steep underbelly of the snowfield, and the rest of this water tower for the West and dance floor for my daughters was, literally, reduced to rubble. The route was now too dangerous for Sol's initiation, so just Atalaya and I skied the dirty snow of the underbelly, squeezing in five laps of five turns each. We tried to laugh at ourselves again for, as always, enjoying whatever we could find. But the laughs were not as hearty.

As an environment and sustainability professor at Western Colorado University, I know the studies well—the winter Elk Mountains could see a seventy-five-day reduction in ski season due to climate impacts by 2090. What does this reduced snowpack mean for ranchers? For forest fires? For the Indigenous homelands of the Southwest downstream? For alpine, subalpine, semi-desert, and riparian species relying on the waters from those snows? For bird migrations through this corridor? For the urban civilizations relying on Western reservoirs? For the peoples of Mexico at the bottom of an unjust water distribution system? These are not abstract thoughts and concerns, and I feel a call to fight against these climate injustices when I contemplate the social and ecological impacts of climate change. But the disappointment in Sol's eyes as she looked down upon the dwindling snowfield of Yule Pass conveyed something more powerful than all of the data on climate chaos I had seen for a quarter-century—I saw loss in her eyes. Grief. It triggered my own visceral sense of loss—loss of a way of life, loss of a way of being, loss of our playful interconnectivity with the very basis of life for all beings. I will never forget that look of loss on Sol's face. It was a look that called me out, that stretched my actions to the reality, livelihood, ecosystem services, and connectivity of unborn generations. It was a look that asked me, "What kind of ancestor do you want to be?"

I first heard this question—*what kind of ancestor do you want to be?*—on the White Earth Reservation of Minnesota while visiting Indigenous leader Winona LaDuke. A member of her community, within seconds of meeting me on LaDuke's back porch, asked me this question and stopped me in my tracks. I wrote about it in *Wildness: Relations of People and Place*:

What kind of ancestor do you want to be?

Michael Dahl's question hangs in the thick Minnesota air. I lean against the deck railing as the chill of evening tightens my skin. Two stories below, Round Lake reddens, matching the maples that are flashing their last burst of fall color. Michael, White Earth Anishinaabe and Midewiwin Lodge leader, offers me the intense grin I suspect he gets when he's just floored a person with an idea. This makes me like Michael Dahl.

"What kind of ancestor do I want to be?" I repeat Michael's query slowly, quietly, as if talking to myself. The question implies that we are, always and already, ancestors. Even before our descendants are born. Even if we never have children. In terms of *space*, nowhere is, ethically speaking, "away." . . . Now Michael asks me to think about ethics in terms of *time*. No era is "away." Ethically, all times and all generations are now. So how are we to live?

Like any effective thinker, Michael posed more questions than answers. I have chewed on this question for over half a decade, and will do so for the rest of my life. Looking at my daughter's disappointment that she would not ski on that day, the question once again floored me.

Returning to the Western Colorado University campus about forty-five miles down-valley from Yule Pass the following week, I visited Alan Wartes at the Think Radio studios to further consider the ancestor question. Alan is a dedicated public intellectual, poet, musician, farmer, and founder of Alan Wartes Media, which comes with a family of podcasts: Think Planet, Think Business, and Think People. He asks elemental questions, and in interviews with him one does not know whether one will be talking about climate data or one's fifth-grade fear of clowns. Walking into his studio, I was greeted by his broad smile; this was going to be a serious day.

After telling Alan that the ancestor question is not about what kind of ancestor we want to *become*, in terms of how we are remembered, but about what kind of ancestor we *already are* (even if we never have kids), Alan shifted into high gear:

WARTES: Well, that's a fascinating distinction. I like that. It's not how do you want to be remembered, which is where your mind might go with that question. . . . . . *Today*, what kind of ancestor do *you* want to be, right now. Today.

HAUSDOERFFER: And there is an existentialist component to it in that the choices you make and the choices you choose not to make are all writing that script. One way that I think about this question, and I am certainly not a physicist, is with that cliché from chaos theory that a flap of a butterfly's wings can set in motion a hurricane somewhere. I see that in terms of each choice. Each choice, each non-choice, is like that flap of a butterfly's wings that leads to a hurricane—socially, ecologically, or morally. We need to understand that. And hurricanes can be resilient disturbances or violent disturbances. They can be good disturbances—prior to overdevelopment and climate change—for mangroves, or they can be horrible disturbances for poorly planned cities in an age of climate disruption. So, how do we take responsibility given that, whether we like it or not and whether we know it or not, each of our choices has that impact on the next generation?

WARTES: Every single one. Cause and effect is a fundamental principle of existence. You cannot escape it. So maybe a corollary to that question would be: What kind of *cause* do you want to be? What kind of effect do you expect your life today to lead to?

HAUSDOERFFER: That's right. Gandhi has this powerful quote that I never fully came to terms with. He says "we cannot obsess over results." Here's someone who produced the ultimate historical result—with his humble body and the basic practice of nonviolence he stopped the British empire. And yet he says you cannot obsess over results. What a powerful paradox to challenge our species for the rest of our time in this universe! He employed the term *Satyagraha*, meaning something like truth force, love force, holding to your truth, holding to your values. In other words, you are not in control of the outcome. I cannot know how I will be remembered. That is why the ancestor question is not about the interpreter of my life long after I am dead; it is about: What values do I want to hold to now? More importantly, it's about: What systems do I want to enliven? Even if I am not remembered, are my values enlivened, and accessible for others to live out, beyond my life? So I think that the ancestor question is a material question.

WARTES: What do you mean by that? I would love an example.

HAUSDOERFFER: Let's go back to Michael Dahl on the White Earth Reservation. He really stumped me. I was there with my questions.

WARTES: You did not have an answer right away?

HAUSDOERFFER: I did not. I did not. In fact, by the time I published the essay in the *Wildness* book, several years later, I still did not have an answer. So Michael shared a story. In talking about their practice of harvesting wild rice on the lakes of Minnesota, he talked about the deep tie to that rice. It is in their creation and migration story, it is in their diet, it supports their economy, it ties them to the ecological health necessary to have wild rice and many other ecological systems for sustainability. So, to me, Michael is blending the material and the spiritual. The spiritual emerges from a relationship with the material. What I mean by that is that for Michael . . . he is living in the place of his ancestors. The soil is enriched by his ancestors' bones, metaphorically and literally. In order to live in that place for hundreds and hundreds of years (they did not have a frontier to move to if they screwed it up), they had to listen to the creator's instructions, spoken through the land, on how to live within the limits and possibilities of that place, so they could co-create the wildness to live on in that place. Then they had to follow their ancestors' instructions for how to live there. So what he's saying is that climate change, acid rain, and corporate economic challenges are threatening their wild rice. And he fears he is not a good ancestor if there is not wild rice for his grandkids, which means he is not a good ancestor if there is not a participation in the struggle to adapt to and slow climate change . . . *so the ancestor question becomes measured by the land.*

WARTES: And measured by the story. As a storyteller, I am fascinated. The story of his people must continue in order for him to be a good ancestor.

HAUSDOERFFER: Yes, and that story is not just told by humans.

WARTES: It's told in the land.

HAUSDOERFFER: Yes, the land is a barometer for how we are living as ancestors, whether or not we are remembered.

Walking out of the Think Radio studio, I reflected on Alan's comments about story being how we measure if we are a good ancestor. Indeed, continuing the right story, in word and deed, measures our ancestral value, and those stories are told in the land. Thinking also about the stories and measurements emerging from Michael Dahl's wild rice, I wondered, "What is my rice?"

As a transient descendent of transient people long-removed from the source of their food, heat, and ancestors: What is my rice? I do not have Michael Dahl's Indigenous, historical connection, nor should I pretend to. But to what extent can I begin to find my own "rice" without co-opting Michael's connection to place and past? In other words, what connects me to the health of the land while also connecting me to the love of my ancestors and to the hopes of my children's children?

As I leave Think Radio studio, my thoughts return to my children, and their (if they so choose) children. I look out at the square blocks of 14,000-foot Uncompahgre Peak to the southwest and see the year's first snows blanketing the high mountains. I smile, hoping that the same storm has begun the process of returning the heart-shaped snowy dance floor to Yule Pass. My thoughts turn to winter, to crisp bright days spent skiing with my daughters. I think of my grandparents, who retired early to a small trailer in New Hampshire so they could ski every day. I think of my mother, a special education teacher and single mother who worked every weekend as a ski instructor so my four siblings and I could ski for free. I think of my formative journey to Colorado twenty-five years ago in search of deeper snow. I think of skiing with my family every winter solstice to our ten-by-twelve cabin, passing on to my daughters the art of the perfect carve through fresh snow. I think of their great-grandchildren and wonder if there will be a seventh generation of ski bums in my family.

I know my rice—it is snowpack. And if that snowpack declines, if my generation does not act radically to mitigate climate change, I worry that I will have failed as an ancestor. Yet regardless of the dire potential outcome of our apathy, acting now as if I already am a good ancestor—connecting my daughters with the power of snow while showing them and my students options for confronting climate change—enlivens a more resilient future for my descendants. Maybe they will even ski in the summer.

ESSAY

Restoring Indigenous Mindfulness within the Commons of Human Consciousness

Jack Loeffler

> Living systems are cognitive systems, and living as a process is a process of cognition. This statement is valid for all organisms, with or without a nervous system.
>
> HUMBERTO MATURANA, *University of Chile*

Nature is a stream whose source is itself, a continuum of seemingly infinite magnitude, a universal commons of unimaginable diversity and complexity. On our planet Earth in our solar system's tiny corner of the cosmos, the phenomenon of life occurred nearly four billion earthly years ago—and with life came cognition. Some scientists contend that life and cognition are two aspects of the same phenomenon, even at the cellular level. If so, the last universal common ancestor of all life, LUCA for short, thought by some to have first occurred as a cell in proximity to a hydrothermal vent in the bottom of an ancient ocean, was imbued with some level of cognition. Thus as life evolved within millions of species through billions of years, so did ancestral re-cognition evolve until it became complex enough to beam into consciousness as experienced within the human and other species.

Just think of that!

We humans are biologically kindred to all of the species that have ever lived on our planetary commons that we call Earth. This is something known intuitively by people whose traditions were shaped by foraging throughout their respective habitats. Those billions of us whose purviews are presently shaped by today's global monoculture largely neglect to honor our common ancestry that extends all the way back in time to our last universal common ancestor. Instead, we have gradually shifted our allegiance to institutions of human provenance such as church, state, and corporation, thus centralizing our cultural perspective. Fortunately, a growing cadre of life scientists is working to restore clarity to our consciousness.

In their provocative book, *A Systems View of Life: A Unifying Vision*, Fritjof Capra and Per Luigi Luisi provide profound insights into the nature of living organisms and their different levels of cognition. I conducted a recorded interview with Capra from which I include the following excerpt.

Fritjof Capra:
[Our book] refers to a conceptual revolution which is related to the conceptual revolution in physics but occurred in the life sciences where people have also discovered that you cannot represent or think of living organisms as machines composed of elementary building blocks, but have to see them as inseparable networks of relationships. We have also discovered that the planet as a whole is a living, self-regulating, self-organizing system. Accordingly, the human body as a machine and the mind as a separate entity have given way to a conception that sees not only the brain but also the immune system, the bodily organs and tissues—in fact, every single cell—as a living cognitive system. So cognition and life have become intimately connected.

Evolution is no longer seen as a competitive struggle for existence but rather as a cooperative dance in which creativity and the constant emergence of novelty are the driving forces. And with the strong emphasis on complexity in the recent two decades—on networks, on patterns of organization—a whole new science of qualities is beginning to emerge.

The big advance of the last thirty years has been to realize that mind is not a thing, it's a process. And the process is called cognition, the process of knowing. Cognition becomes closely associated with the concept of self-organization of living networks. Cognition is actually the process of self-organization of life, so that when you have a cell, the simplest living organism, then the constant generation and regeneration of the cellular components—the process of self-organization—determines how the cell functions and lives. It's not determined by outside forces. The basic rules of interactions are set up by the system itself. That's why we call it self-organizing. And the process of self-organization is cognition. It's a mental process. It's part of the natural world. And the best we can do is to learn from the wisdom of Nature because the wisdom of Nature is not just a beautiful metaphor, but is actual wisdom that is millions and billions of years old. And evolution, although not having a plan and a strategy, nevertheless has been able to tinker with various solutions to its problems, and over billions of years has found the best solutions.

And so this whole new movement of eco-design, of deriving design principles from the basic principles of ecology that we observe in Nature, and the practice of bio-mimicry, of mimicking the solutions that Nature has found, is something very important. Indigenous cultures have had this approach for centuries, gleaning things from Nature, copying things from Nature, and living in harmony with Nature and cooperating with Nature.

Native cultures have been engaged in indigenous mindfulness for the duration of our species. We attained species-hood some two hundred thousand and more years ago during the Pleistocene epoch that lasted for two and a half million years and ended around twelve millennia past. Most of our time as a species was shaped within the last great Ice Age.

We've continually adapted to the circumstances of our respective habitats, and over the course of ten thousand generations of ancestry, have coevolved into what we are today. Twentieth-century ecologist Paul Shepard forwarded the theory that we were shaped as a species genome-wise as Pleistocene hunter-gatherers. "We are free to create culture as we wish, but the prototype to which the genome is accustomed is Pleistocene society."[1]

Since the end of the Pleistocene, there have been enormous cultural shifts thought to have originated in part with the advent of the development and practice of agriculture sometime around 9,500 and 11,500 years ago in the Levant and China. Some estimate there to have been two to ten million humans on the planet, few of whom had yet to arrive in the Western Hemisphere.

While the whole world is a commons, the human species rapidly began to claim first dibs on entire habitats. Humans began to migrate across the Bering Landmass and gradually populate the Western Hemisphere. By 1492 CE, an estimated 75,000,000 humans lived throughout North and South America, where as many as two thousand distinct languages were spoken.[2] Two hundred fifty of those languages were spoken in North America and three hundred fifty more in Central America. The rest were spoken throughout South America.[3]

These Native Americans had culturally evolved in vastly diverse habitats that extended from sea level to mountain peaks that rose to above 20,000 feet. From jungles to deserts to plains to forests, the Western Hemisphere was spectacularly rich in flora and fauna. However, during the five most recent centuries, the sons and daughters of Europe decimated the native human populations and wrought horrendous damage to these once-thriving habitats. The institutions of church, state, and corporation have all but rung the death knell for Indigenous peoples and many of their habitats.

Up to six and a half million Native Americans presently remain in the United States, some of whom continue to speak their native languages, and live their lives in accord with their traditional cultural and spiritual coordinates.

The late Rina Swentzell was born into the Tewa village of Kha'po Owingeh, known to outsiders as the Santa Clara Pueblo situated along a western bank of the northern Río Grande. She addressed her sense of the earthly commons as understood by many Puebloan Indians.

Rina Swentzell:
From a Pueblo point of view, the commons is everything. It is the context that we live in. In the old Pueblo thinking, the community was always thought of as being whole. Everything was interconnected. There was always a center to it as well, and I was a center and you were a center. There were many centers as a part of the whole. With the Pueblo people, there are so many things that happen simultaneously. The wind is blowing, the water's flowing, and we're actually walking around and talking. It's all part of this idea of what we all share. It's that notion of sharing.

In that Pueblo context, the focus was always, what is it that surrounds me? Who and what surrounds me? We felt that it was the earth, the sky, the clouds, the wind, and that incredible term that we have that says it all: it's the *p'o-wa-ha*—it's the water-wind-breath, the thing that we're feeling right now. It moves through our entire world in such a way that it connects everybody and everything. That becomes the commons. What is it that's blowing through the window right now that's giving us all vitality actually? That's the flow of life. In the Pueblo, it really was that thing that swirls around, that swirls, that moves, that creates that sense of commons because it's the ultimate of what is common to every living being. What do we have in common with the trees, with the rocks, with everything that makes our life what it is today?

The planetary commons that sustains all life comprises countless ecosystems. Each ecosystem is a self-regenerating system that includes all species cooperating within a geophysical habitat. The human species is but one of millions of species that might inhabit any ecosystem. Native peoples who still adhere to their respective traditional values intuitively recognize this systemic interconnectedness and thus honor their fellow species—plant, animal and otherwise—the true source of their ancestral wisdom, the commons of human consciousness.

Philip Tuwaletstiwa is a citizen of the Hopi Independent Nation situated in the heart of the Colorado Plateau. He is now retired as assistant director of the United States Geological Survey. His knowledge of homeland extends far beyond the parameters of transits, theodolites, and tripods.

Philip Tuwaletstiwa:
We talk about the *Sipapu* [place of Hopi emergence] as being a place that is always in our memories, our thoughts. It is only one of thousands of places like that. There are hundreds of shrines all over northern Arizona, and southern Utah, and southern Colorado that are connected in our minds and our consciousness. We think of them in terms of "what do they mean to us as a tribal people?" We know that this clan, that clan was there. We know that this is the place where you go to get a particular mineral, an herb, a particular plant. We are familiar that something happened there a long time ago that affected us. We are emotionally connected to it, and that is why Hopis are emotionally connected to our landscape in its entirety. We can articulate that connection to hundreds, if not thousands of points on the landscape. It is like a spider web connected to all of these things. And the Hopis are connected to the spider web. We are all interconnected. And we cover this ground up here in our consciousness, in our subconscious, in our culture, in our language.

The geophysical cradle of any ecosystem is fundamental. Mountains, rock formations, watersheds, the terrain itself are all embedded in the deep consciousness of Indigenous peoples. Place plays a mighty role in fomenting Indigenous mindfulness and defining cultural attitudes. The nature of homeland shapes cultural perspective and is thus, in the most profound sense, our terrestrial ancestor.

The great cultural anthropologist Edward T. Hall well understood the important intersection of cultural and biological diversity. He had both an intuitive and intellectual understanding of the role of cultural consciousness within the context of respective homeland.

Edward T. Hall:
The land and the community are associated with each other. And the reason that people get their feeling of community from the land is that they all share in the land.

Ethnicity is one of the greatest resources that we have in the world today. What we have here are stored solutions to common human problems, and no one solution is ever going to work for over a long period time, so we need multiple solutions for these problems.

Culture is an extension of the genetic code. In other words, we are part of Nature ourselves. And one of the rules of Nature is that in order to have a stable environment, you have to have one that is extraordinarily rich and diverse. If you get it too refined, it becomes more vulnerable. So we need diversity in order to have insurance for the future. You need multiple solutions to common problems. The evolution of our species really depends on not developing our technology but developing our spirits or our souls. The fact is that Nature is so extraordinarily complex that you can look at it from multiple dimensions, and come up with very different answers, and each one of them will be true. And we need all of those truths.

These truths become the basis for the commons of human consciousness. A profound concept to have re-emerged from the counterculture movement of the 1960s and '70s is *bioregionalism*. A powerful proponent of bioregionalism is poet and environmental philosopher Gary Snyder. One July morning in 1985, while sitting in a shady pine forest in the Sierra Nevada foothills, I recorded Gary Snyder as he spoke his definition.

Gary Snyder:
"Bioregion," the term itself, would refer to a region that is defined by its plant and animal characteristics, its life zone characteristics that flow from soil and climate—the territory of Douglas fir, or the region of coastal redwoods, short grass prairie, medium grass prairie, and tall grass prairie; high desert and low desert. Those could be or verge on bioregional definitions. When you get more specific, you might say Northern Plains short grass prairie, upper Missouri watershed, or some specific watershed of the upper Missouri. The criteria are flexible, but even though the boundaries and delineations can vary according to your criteria, there is roughly something we all agree upon. Just like we agree on what a given language is, even though languages are fluid in their dialects. So bioregionalism is kind of a creative branch of the environmental movement that strives to re-achieve indigeneity, re-achieve aboriginality, by learning about the place and what really goes on there.

Bioregionalism goes beyond simple geography or biology by its cultural concern, its human concern. It is to know not only the plants and animals of a place, but also the cultural information of how people live there—the ones who know how to do it. Knowing the deeper mythic, spiritual archetypal implications of a fir, or a coyote, or a blue jay might be to know from both inside and outside what the total implications of a place are. So it becomes not only a study of place, but a study of psyche in place. That's what makes it so interesting. In a way, it seems to me that it's the first truly concrete step that has been taken since Kropotkin in stating how we decentralize ourselves after the twentieth century.

The practice of bioregionalism is the restoring of indigenous mindfulness within the commons of human consciousness.

We can't go home to the Pleistocene. That epoch ended nearly twelve thousand years ago, and for the last several millennia, our species has engaged in a spasm of cultural evolution that has spawned enormous population growth, rearranged our collective perspective, and forever altered the nature of the planetary biotic community. We have left our Edenic Pleistocene heritage to pursue a different sort of destiny that now largely excludes recognition of our place in Nature. Agriculture, the Industrial Revolution, science and technology, modern media and cyber-culture, transportation, religious mores, corporate economics, and global politics are among the myriad factors that dominate our current collective consciousness. It is interesting to consider that of the one hundred six billion humans estimated to have lived on our planet Earth since we attained full species-hood tens of thousands of years ago, seven and a half billion or roughly 6 percent of us are presently alive. Before the end of the Pleistocene, the human population of planet Earth is thought to have fluctuated between two and ten million. And for most of our tenure on this planet, none of us inhabited the Western Hemisphere, where the flora and fauna engaged in evolving levels of reciprocal cognition and self-regeneration without human input.

We now recognize ourselves as the keystone species and thus should be ethically compelled to proceed responsibly. Instead, we have become dominated by self-obsession rather than the well-being of our biosphere. We neglect diversity of perspective. If we continue from this centrist position, we doom to extinction both ourselves and thousands of other species, kindred through common ancestry with whom we have coevolved through the billennia. Shall we so trivialize our consciousness?

The planet will survive, life will survive, but human consciousness may not survive. Through our hubris, we may close a window through which the universe may observe itself. From the point of view of the universe, the loss of human consciousness may seem infinitesimally small. From our human perspective, the evolution of our species might well be but a complex biological/cognitive experiment that fails by our inability to survive our arrogance.

My friend Camillus Lopez is a Tohono O'odham lore-master who lives in the heart of the Sonoran Desert bioregion. His people inhabit their traditional homeland that is now sundered by the international boundary with Mexico.

Camillus is intent on maintaining his traditional perspective and ensuring that it does not disappear from the cultural consciousness of his people.

Camillus Lopez:
Community is everything. It's the stars. It's the ground way under. It's the little ant that comes across. It's Coyote. It's the buzzard. Your actions reflect who you are. And if you can see yourself in it, then you're there. *But if you can't look at Nature and see yourself in it, then you're too far away.* That's why I think one of the things people need to do is go out and look at the mirror of Nature and try to see themselves in it. Because if they can see themselves in it, then they can help themselves by helping the environment.

*

[All of the excerpted spoken interviews were conducted by Jack Loeffler and are located in the Loeffler Aural History Archive at the New Mexico History Museum at the Palace of the Governors in Santa Fe.]

ESSAY

Reading Records with Estella Leopold

Curt Meine

Estella Leopold and I meet along the Wisconsin River on a drizzly fall day, on the land that her father, conservationist Aldo Leopold, depicted in his classic book *A Sand County Almanac*. Over our thirty-five-year friendship, Estella and I have had many opportunities to talk over events in her life and in the lives of her family, and the insights that she has gained along the way. But when I pose the question to her—"What kind of ancestor do you want to be?"—I am not sure how she will respond. She pauses for just a moment.

Estella was the youngest, and is now the last survivor, of five notable siblings—the children of Aldo and his wife, also named Estella. "Little" Estella is now ninety-one years old. She was just eight in 1935, when her father acquired, for a song, the "sand farm in Wisconsin, first worn out and then abandoned by our bigger-and-better society." She spent her teen years actively engaged in the pioneering ecological restoration work that began to heal the damaged land. She was twenty-one in April 1948, when her father lost his life just yards away from where we are sitting. He succumbed to a heart attack that overcame him while fighting a grass fire that had escaped from the neighbor's farm.

Estella studied botany at the University of Wisconsin, then earned a master's degree at the University of California, Berkeley and a PhD at Yale. Her research involved early work in the emerging field of palynology—the study of fossil pollens and spores to understand plant evolution, ancient ecosystems, and past climatic conditions. Her significant contributions to paleobotany earned her multiple honors, including election to the US National Academy of Sciences. (Her brothers Starker and Luna were also elected members, a unique accomplishment in the history of the Academy.)

For all her achievements as a scientist, Estella was never one to wall off her research from her conservation activism and advocacy. Her knowledge of the deep past was prelude to her concern for the future. In the 1960s she led efforts to establish the Florissant Fossil Beds National Monument in Colorado. She fought proposals to construct dams along the Colorado River that would have flooded portions of the Grand Canyon. She opposed plans to bury high-level nuclear wastes along the Columbia River at Hanford, Washington. She helped create the Mount St. Helens National Volcanic Monument in 1982. With her siblings she founded the Aldo Leopold Foundation to carry forward her father's "land ethic," his philosophy of love and respect for land "as a community to which we belong."

Given Estella's informed scientific understanding of deep time, and her family's legacy of conservation and ethics, I expected, I suppose, an expansive response to the matter of responsible ancestry. I reframe the question, and ask her what comes to mind when I invoke the term *ancestor*.

She has a one-word reply.

"Records!"

Some explanation is in order. Estella is working on another book. Since her older sister Nina passed away in 2011, Estella has devoted more of her time to writing about her personal and family history. In 2012 she published *Saved in Time*, recounting the campaign to protect the Florissant Fossil Beds. In 2016 she shared the experience of her family's intimate practice of land stewardship in *Stories from the Leopold Shack: Sand County Revisited*. Now she is researching her next writing project, focused on her mother's deep family history in New Mexico. Her thinking on ancestry is thus quite immediate, literal, and personal.

"We need to keep records out of respect! They give a feel for what our elders were like." The stories of her Luna and Otero ancestors are woven into and through the history of Mexico and New Mexico, embedded in place names across the region. Her ancestor-patriarch Solomon Luna, for example. He was a sheep baron whom Aldo once described as "kind of a king-maker to New Mexico," playing a key political role as New Mexico gained statehood. Other stories involving other maternal ancestors reflect enduring family traits: warm familial relations, a tradition of gracious hospitality, an abiding love of music and Mexican cuisine, a steady commitment to progressive political reforms. In her nineties, Estella is still exploring who she is and where she came from. And to have *records*—accounts, interviews, stories—is to have *data*, real information to help illuminate the unknown.

Maybe Estella has in mind one of her father's lines. Reflecting on his lifelong interest in phenology, the careful chronicling of the timing of natural events,

Aldo wrote that "keeping records enhances the pleasure of the search and the chance of finding order and meaning in these events." Discovering order amid the fog of time and change, details and impressions. Finding meaning in the layers of natural and human history. Estella built a long and distinguished scientific career by seeing whole worlds and geological epochs in microscopic fossilized grains of pollen. She knows how to make the most of the minutest bits of evidence. It is what she was trained to do.

But Estella moves easily from the hard nuggets of facts and evidence to the fine latticework of our constructed ethical norms. For her, being a responsible ancestor is not an overly complicated matter. She says simply: "You have to be pretty moral, and set a good example." The world knows Aldo Leopold for his framing of an inclusive land ethic. Estella sees its germ in her parents' lived experience and character. "They were kind to everyone, both of them so generous. They were so in love with one another, and lived in a way that transferred and passed along that kindness." Her mother was universally appreciated for her personal warmth and thoughtful consideration of others. Her father was "always looking for the highest qualities in all his acquaintances—his classmates, his colleagues, his students." From this relationship came a family ethic of careful regard and respect for other people. That, in turn, yielded—or maybe coevolved with?—a land ethic, extending that regard and respect to the larger community of life.

I wonder if Estella can draw a connection between her refined scientific understanding of deep geological time and her sense of responsibility extending across generations. I coax her to connect her knowledge of our fossil ancestors to our obligations to those who will follow us. "Estella, do you think having that view of deep time has sharpened in some way your perspective on our ethical commitments?"

She again resists thinking conceptually about the question, and goes straight to a story. She recollects the time she invited the famed Wyoming-based geologist David Love to one of her study sites in the coastal Mexican state of Nayarit. Having explored in depth the Eocene flora of Wyoming, she tempted Love by explaining that the ancient plant life of Wyoming was best represented in contemporary times by the tropical forests of Nayarit. "Come to Mexico, David, and you can imagine what Wyoming was like! You can see it there." Her science feeds her imagination. When pursued in this way, empirical knowledge allows us to put our brief human experience on Earth—as individuals and as a species—into context. To imagine ourselves in our most expansive context of time and space.

In the 1930s Aldo Leopold found himself also in Mexican space, navigating the river of time and expanding his imagination. In Chihuahua's Sierra Madre, along the banks of the Rio Gavilan, Leopold heard "the song of the river":

> To hear even a few notes of it you must first live here for a long time, and you must know the speech of hills and rivers. Then on a still night, when the campfire is low and the Pleiades have climbed over the rimrocks, sit quietly and listen for a wolf to howl, and think hard of everything you have seen and tried to understand. Then you may hear it—a vast pulsing harmony—its score inscribed on a thousand hills, its notes the lives and deaths of plants and animals, its rhythms spanning the seconds and the centuries.

For all who listen to the songs of rivers, or are attentive to pollen grains, or confront honestly the mixed stories of our own ancestors, the records of the past help us to peer into the indefinite future. The messages from history cannot on their own make us into the ancestors we would like to be for our descendants. But they help us to imagine ourselves as honored bearers of stories and responsibilities, and they bond us to those who will follow.

As we talk, Estella Leopold draws on a past that reveals itself to her at the end of a microscope, in the pages of family documents passed along by her mother, through memories of walking the sandy shores of the Wisconsin River with her father. All that history distills itself, finally, into an ethic of care that is simply a *given*, a natural inheritance for her. She continues to pass it along, to give it forward. Because you have to be pretty moral. You have to set a good example.

ESSAY

How to Be Better Ancestors
Winona LaDuke

> How long are you going to let others determine the future for your children? Are we not warriors? When our ancestors went to battle they didn't know what the consequences would be, all they knew is that if they did nothing, things would not go well for their children. Do not operate out of a place of fear, operate out of hope. Because with hope anything is possible. . . . .
>
> THUNDER VALLEY COMMUNITY DEVELOPMENT

I believe in place. *Anishinaabe Akiing*, the Land to which the people belong, that's where I live. I live in the same area as my great-great-great-great-grandparents lived. *Nimanoominike*, I harvest wild rice on the same lakes, canoe to the same berry patches. I am eternally grateful to my ancestors for their consistency and their commitment to land, to ceremony, and to those who had not yet arrived, like myself. My lake, Round Lake, is where the so-called last Indian uprising in Minnesota occurred. And I am eternally grateful to the Skip in the Day Family for demanding justice on our Lake and for stopping the timber barons from stealing all of our great and majestic pines. In walking, riding a horse, or canoeing these lakes and this place, I remember those ancestors. And I offer them food and prayers. Those are cool ancestors, great role models.

My mother is the artist Betty LaDuke, and my family on her side hails from Ukraine. They were Jewish farmers who became union workers in New York City. My great-great-grandfather had a windmill to grind wheat, and was displaced by the burning of coal, and the progress of new mills. My grandmother worked in the garment district and my grandfather worked as a house painter. Decent people. Courageous people. Humble people. I feel that I not only remember them, but live their lives in my own way, particularly that transition of power my ancestors experienced, from their way of life working with water, working with wind, to this fossil fuels mess that I'd like to reverse.

And so, as I reflect on the question of how to be a good ancestor, I reflect on intergenerational accountability. How do I account my behaviors and decisions to my ancestors and to my descendants?

This is easier for some of us than others. America does not remain in Anishinaabe Akiing. Privileged by the fossil fuel economy, which has amplified and intensified our disconnections, we are transient, we move. Few people live in the same place as their ancestors, and many more of us have historical amnesia. Perhaps this particular amnesia is learned as a coping mechanism, or perhaps it is a consequence of the segregation into not only an anthropocentric, but an increasingly self-centered worldview, that aggravates this condition. Not knowing history has huge perils. Ecological amnesia is when we forget what was there; transience complicates all of this.

Transience means that we do not come to know and love a place. We move on, and in so doing are not accountable to that place. Always looking for greener pastures, a new frontier, we, I fear, lose depth, and a place loses its humans who would sing to it, gather the precious berries, make clean the paths, and protect the waters. My counsel is stay: make this place your home, and defend this land like a patriot.

I look to *mino bimaatisiiwin*—the excellent life offered to the Anishinaabeg by the Creator. In this life, the basic teachings are elegant and resonate: care for yourself, the land, and your relatives. Remember that this world is full of spirit and life and must be reckoned with. The land of berries, wild rice, maple syrup, and medicines comes with a covenant, an agreement between the Anishinaabeg, or myself, and the Creator. Keep that covenant, that agreement that we will take care of what is given to us, and your descendants will be grateful.

Understand your responsibility for this moment. I understand mine. As I watch my brothers and sisters to the west at Standing Rock protect the water in the face of rubber bullets, tear gas, and the spraying of poisons, I am awed and inspired, and remember that I am one of them. In this moment, not unlike the Selma moment, be present.

Standing Rock is not only a place, it is a state of mind, it is a thought, and it is an action. In a time when the rights of corporations override the rights of humans, stay human, and remember that the law must be changed. For civil society is made, as democracy is made, by the hands of people, courageous people, and is not a spectator sport. While at one time slavery was legal, it is no longer, and soon we must free our Mother Earth from her slavery to an exploitive economy and ensure her rights.

In each day there is a heartbreak of story, a constant heartache for our relatives, whether they have wings, fins, roots, paws, or hands . . . but there is much beauty and joy in the midst of heartache. Hold your sorrow and grief, remember, but be grateful for this life. The Creator has given us a good one. And your descendants will be grateful for this good life, this Minobimaatisiiwin.

In this time, do not underestimate yourself, nor the power of what is larger. As we saw at Standing Rock, unity, hope, a worldwide outpouring of love and support emboldens water protectors worldwide and that is something we will all need, along with our Mother. How that power is actualized is up to each of us, but acknowledging our responsibility for power is how we are accountable intergenerationally.

Two lessons I take from one of my great teachers, Wes Jackson of the Land Institute. As you contemplate your choices, mill about. This is to say, that if you can live in your one acre, do so, mill about on that one acre, and do not move. Perhaps that lesson is live simply and care for the place you know so that those who follow can live there, too. And, believe. Wes said one time that if you're working on something that you plan on finishing in your lifetime, you are not thinking big enough. Let us use the gift of our thoughts, and in the words of the Great Hunkpapa leader, Sitting Bull, "Let us put our minds together to see what kind of future we can make for our children...." Then we will be great ancestors.

INTERVIEW

Wes Jackson

John Hausdoerffer and Julianne Lutz Warren

HAUSDOERFFER: *So, Wes, what kind of ancestor do you want to be?*

JACKSON: I hope to be among those members that made conceptual and material changes in the course of history, which is to say help move humanity from an extractive to a renewable economy. We have had great gains. We have become the first species on this planet to become matter's and energy's way of having gained self-recognition. We have great knowledge of our journey from the Big Bang to Darwinian evolution. We have a greater sense of the journey from minerals to cells. Our scientifically verifiable cosmology is different than the ancient cosmologies. With that recent cosmology, one hopes it will inspire us in the arts, in music, in science, in our political life as well, and help us move from our extractive ways to living within our means. That is a huge challenge, of course, and it will have to be, more or less, our journey from now on if we are to stop soil erosion, chemical contamination of our land and waters, putting more carbon in the atmosphere, and the existential threat of nuclear weapons.

WARREN: *I think leading from that, is a good place to ask, when you said "we," who do you mean?*

JACKSON: *Homo sapiens.*

WARREN: *To be more specific, how might you see your work in relationship with or as a response to colonialism?*

JACKSON: Americans are still colonialists. Colonialists do not "discover." They want to know how the world works, in order to extract from it. To become native, we need to understand more of how the world is. "Is" and "works" are, of course, always intertwined, but to make that transition from regarding the Earth as a mine or quarry to seeing it as a source of renewal will require more interest in how the world "is." Then we will become native to our places.

HAUSDOERFFER: *A creative source. Not just a source. Because "source" still sounds like that "mine or quarry" you just mentioned.*

JACKSON: The Earth as our creator, our maker, and defender. (The upper atmosphere protects the lower biosphere and makes it safe for life.) With proper restoration, the Earth is our redeemer, as we redeem it.

HAUSDOERFFER: *The buffalo were out the other day when we did a tour of the Wauhob Prairie at The Land Institute. I think of that prairie and the biodiversity of the prairie grasses and flowers not as so much "the Earth." I think of them instead as deeply rooted homelands of Indigenous Peoples who rely on the buffalo to be who they are. I think of those Peoples' choice to burn, their choice to allow for the buffalo to live with a certain amount of wildness, to stir up the soil and fertilize it. I'm curious, how do you see the Native inhabitants of this place as your ancestors? And what are your responsibilities to their descendants?*

JACKSON: Before the entrada, the natives lived on a sun-powered world with no fossil carbon. Their patterns were in conformity to that world. They are our ancestors in an important sense, but I don't know if I can speak to their descendants.

WARREN: *If I could just enlarge the idea of ancestor to peoples' identities and culture. The people who identify and still carry forward and have saved their cultures through all of the trauma. Those people. Those descendants who are still here.*

JACKSON: So, what would that future look like? The lifeways of Indigenous People are a kind of standard, a source of mindfulness, a way not likely to be repeated except in small ways. The landscapes have changed and a new cultural handing-down will come. There will be a transition, with our eye always on that future necessity to live without nuclear weapons and an increase in greenhouse gases. It will be a tough journey. Even The Land Institute's research features industrial equipment. We have several tractors, pick-up trucks, three greenhouses, a fancy lab, lots of fossil fuel–dependent equipment, millions of dollars of investment in our fossil fuel–dependent research. We use modern scientific knowledge to bring perennial grains and polycultures into existence. Here's a question: Will the technology that brought these new varieties into existence be required to sustain them? My answer is: no. Their creatureliness will sustain them, and would have been available even to a Neolithic gatherer and hunter. The industrial world can't say that about their renewable technology, whether it's wind machines or solar collectors or Priuses or squiggly light bulbs.

This is to say, if we have to walk the talk, we'll never get there. As a part of the great transformation, we will utilize much that brought us here, and maybe even ramp it up some. The sequencing of the Kernza genome is pretty high-tech. But consider the rich and new insights derived from this high-energy era. Three of our scientists have published papers that explain why our ancestors never developed perennial grains and why it seems we can now. That required new insights in genetics, and modern computational power. The scientific revolution and fossil fuel is behind all of that.

What about the First Nations people and some of their patterns on the land within a sun-powered-only world? We can only imagine what those patterns were like. Like us, they evolved in tribes. They had traditions that can serve as reference points. Their traditional beliefs held that the Earth is our mother. Our verifiable history confirms that. With a little stretch, the cosmologies of many ancient people must have been somewhat isomorphic with ours. There will have to be fewer people overall, of course.

HAUSDOERFFER: *Is it fair to say one way of thinking about how Native inhabitants of your prairie are ancestors and how you're an ancestor to their descendants, people who share that cultural identity, is that there's a shared co-creation of wildness? Just as they had social choices to burn grass and follow the buffalo in a way that enriched the prairie, so you have the social choice to develop perennial polyculture in a way that will result in enriching the function of the prairie. Even though these are two different ways of being on one piece of land, in both cases there is co-creation in the sense that the production of livelihood leads to more biodiversity. Is that a shared ancestral legacy? Do you think of your work in that co-creative way?*

JACKSON: I see where you are going. I think that long before we become successful in our ideals, we may want to drop the terms "domestic" and "wild." When we look out on our native prairie, there are countless domesticities. In aggregate we call that prairie ecosystem "wild." A concrete bridge over a stream is a domestic product of ours. A cliff swallow who looks at a human-made bridge doesn't seem to care whether it's a domestic product or a natural cliff. What we are really after are processes. We hope and expect our herbaceous, perennial, seed-producing polycultures to bring the processes of what we now call the wild to the farm landscape. The specific species of the former wild won't be there, but their analogues will. When I say "wild," I especially mean the necessary processes. If we are to eat grain, we need the grain-producing analogs.

HAUSDOERFFER: *I appreciate that point, thank you. Even if we do away with dichotomies like "domestic and wild," how do you react to my suggestion that there are parallels between how perennial polyculture and Indigenous lifeways both co-create ecological health and human livelihood simultaneously?*

JACKSON: As I just mentioned, I do expect perennial polycultures to return at least most of Nature's processes to the farm. By "mimicking the structure we hope to be granted the function," as my friend, Professor Jack Ewell, said of his tropical rainforest experiments. Our local geniuses are the prairies, which have warm and cool season grasses, legumes and members of the sunflower family. We call those four "functional groups." If we can mimic that structure, those ecological "slabs of space-time" will be healthy. They save soil from being lost. They feature nutrient recycling, and along with species diversity they have chemical diversity, which means that it would take a tremendous enzyme system on the part of an insect or pathogen to trigger an epidemic. That amounts to ecological health at the ecosystem level. Indigenous lifeways running on contemporary sunlight were intimately plugged into the ecosystems that supported them. Perhaps by starting with the integrity of Nature's ecosystem, we will have the standard for an ecological view compelling enough to replace the industrial mind.

WARREN: *Do you want to say anything more about responsibilities you maybe feel towards current descendants of Indigenous Peoples of the prairie, like the Kanza People, the Kaw Nation?*

JACKSON: We need to save all of the wildness, including cultural wildness. No matter that most of those "without-fossil-fuels" cultures are gone, Indigenous peoples carry some of the cultural memory of the pre-fossil-fuel world. We need the useful abstractions for the future we want.

WARREN: *Let's take some scientific evidence that you've mentioned recently, that is, evidence that points to Earth's life originating from chemicals mixing around ocean vents, and what you take from that. Why do you feel that is so important? I think that also might touch on, or if you'd be willing to touch on, what that importance might have to do with pre-Baconian, and different also contemporary ways of knowing? Because, what I heard you saying, and from knowing you, is that verifiable information is very important. Then, too, there is this ancient history of Peoples around the world, for example, already considering stars as their ancestors. So, this all seems to come together in this question of what this very recent evidence means to you that life first happened in the ocean?*

JACKSON: It is an important part of our story, but you are asking, "Why do we need to know all of that?"

WARREN: *Yes, what does it mean to you?*

JACKSON: Perhaps it comes down to where do we stand on the creation. I was raised to be a Christian, and am now what I call a five-eighths Christian, a "Sermon-on-the-Mount believer" and a "he-who-is-without-sin" sort of Christian (it has been said there is still Methodist in my madness). A new source of purpose has replaced much of the earlier Christian purpose with our new cosmology. It is interesting how a modern Jew, the late Nobel laureate biologist George Wald, put it. He said something like, "We living things are the late outgrowth of the metabolism of our galaxy. The carbon in our bodies was cooked in the remote past of a dying star and from it, at lower temperatures, came oxygen and nitrogen which were spewed into space to make planets and all of life." He continued by saying, "The ancient seas set the pattern of ions in our blood, and the ancient atmospheres molded our metabolism." Knowing all of this tells me that we are tuned to our creator Earth, not here to please a man-god.

HAUSDOERFFER: *Is it important to call that dying star an ancestor?*

JACKSON: It is an ancestor.

HAUSDOERFFER: *But is it important?*

JACKSON: Is it important? It's verifiable. It's one of the steps in our ancestry.

HAUSDOERFFER: *And what can a dying star teach us about how to live?*

JACKSON: That star was the early part of our journey. It seems we have been cycled through a supernova at least twice. Stars are born, grow, age, and die, and materials are returned. Knowing all of this and how we came to be on Earth should teach us that we cannot live on Mars. Consider the difference in gravity, and how our developing, and even adult, bodies are keyed to a constant gravity around the world. Our metabolism was molded here, not on Mars. The nitrogen/oxygen ratio we have evolved with here will not be found on Mars. That's just one very practical consideration. The goals of the Mary Evelyn Tucker/ Brian Swimme "Journey of the Universe" project, in part, are to help us appreciate that the universe is "not a place, but a story." As a story, they hope it will help all of us as members of a "meaning-making" species in our work for a better world. Does that answer your question?

HAUSDOERFFER: *It does, and a follow-up would be: if perennial polyculture is successful in the way you envision it, how might people five hundred years from now view perennial polyculture as an ancestor? How do you hope they see it as an ancestor?*

JACKSON: It will be one of the ancestors. How might perennial polyculture be viewed five hundred years from now? Well, corn and sorghum fields look very differently than when I was a kid. Kernza is being changed faster that I thought it would. The largest transformation will be cultural, whether by design, nuclear holocaust, global warming, eroded soils, famines. How can we know? So much can happen in a short period of time. One of my grandmothers shook the hand of President Lincoln at Gettysburg in November of 1863 and held my hand many times between 1936 and 1938. I hope that soil erosion and chemical contamination of the land and waters will have stopped. Perennial polycultures, we imagine, will play a major role in all of that. A cultural revolution will have to be a part of it all.

HAUSDOERFFER: *So, it's an ancestor in a larger system yet to be determined. Is that you as an ancestor? An ancestor in a relational system?*

JACKSON: Well, we see ancestorhood because of our limited time perspective.

HAUSDOERFFER: *So, let's redefine it.*

JACKSON: I like knowing what Hans Bethe said in his Nobel acceptance speech. Bethe discovered how our Sun works. In his laureate's talk, he mentioned that stars are born, grow old, die, and the materials are returned to do it all over again. We've been cycled through a supernova at least twice,

I understand it is only a story, but it enlarges our part of the brain given to pondering such questions. When someone asks why is the Earth round, that someone may be a prime candidate to help our thinking on how we can pull out of this nosedive we are in as part of our story.

WARREN: *That's amazing.*

JACKSON: It begins by acknowledging that the universe is not an object, but a story.

WARREN: *That's beautiful. It's amazing what's emerging from the work that you're doing. I wonder if I say to you that work is of its own, it's new, and yet, some of the ideas that you're coming to by studying and consulting with the genius of the land for all these decades bring you right back to some starting places where some Indigenous Peoples have already gleaned knowledges and then have saved for thousands of years in their own ways.*

JACKSON: We do have that in common. As we have been saying, Indigenous Peoples lived in a solar-powered world. Many of us have as a goal, as an eventual necessity, to live in a solar-powered world. We want that centralized nuclear reactor [the Sun] to continue to deliver photons on a trip that requires only eight-and-a-half minutes to be captured by a chlorophyll molecule. That is nothing short of amazing. In our time, we have to begin the journey to "get our act together" in order to live on contemporary sunlight. All of us had ancestors that ran on contemporary sunlight. They probably all had creation stories—The Epic of Gilgamesh, The Garden of Eden, and the Native Americans. There seems to be consistency here and there among the different cultures in that solar-powered world. We now have another story that is not contradictory in the main. Our scientifically verifiable story would not be here without the highly-dense carbon era.

WARREN: *And "we" being?*

JACKSON: "We" human beings.

WARREN: *The colonizers that have come and are just now figuring some of this stuff out?*

JACKSON: To be fair, the Native Americans were not really tested with what they would have done if they had the technology and the fossil fuels. Had they been tested by the 3.45-billion-year-old carbon imperative, I suspect they too would have gone for it.

WARREN: *I wonder, though. Why did that technology get invented in the first place, and why by some groups of Homo sapiens and not others even in areas with coal and oil deposits? I believe that we've got to be really honest, and really very careful about a lot of stuff here. And, this warrants much bigger, deeper conversations that involve various Native Peoples speaking for themselves.*

JACKSON: This is an important question that needs some elaboration. It brings to mind the Jefferson Building, which is the Library of Congress, in Washington, DC. It may be the most beautifully decorated building in our country. Inside, looking up, we are able to get some sense of who we modern citizens are as products from other places and periods. In the dome of the Main Reading Room, which is the most central and the highest point, there is a reminder of the better part of our journey in civilization. The genocide of Indigenous Peoples is not there, nor is the legal chattel slavery which our original Constitution allowed. Let's not keep those two realities too far aside. We see the twelve figures in chronological order representing our journey. Egypt represents written records, with the hieroglyphics of the seal of Mena, the first Egyptian King. Then comes Judea, which gave us the dominant religion. In Hebrew characters on the face of the pillar is: "Thou shalt love thy neighbor as thyself" (Leviticus 19: 18). Next is the reminder that the Greeks gave us philosophy, Rome administration, Islamic peoples physics, the Middle Ages our modern languages, Italy the fine arts, Germany the art of painting, Spain discovery, England literature, France emancipation, and finally the figure for America, science. This is clearly a stretch, in that the Royal Society was formed in England in 1660. Never mind that an electric dynamo represents our contribution to the advancement of electricity. That is a small part of the origins of science.

We can be proud of all that, but here is my point. No group of people has either spiritual or genetic claims to innate superiority in intellectual or moral realms. In our time we have to challenge American exceptionalism, which is to say the United States has special status as the moral exemplar of the world, along with European exceptionalism. Sure, we are mostly products of Europe culturally, but Europe's domination of the world was not the product of intellectual or moral superiority. And now I must say it: it follows that we must reject any culture's claim to be exceptional. Human cultural variation is a product of geography, climate, and environmental conditions, not divine intervention or genetics.

I mentioned geography. Eleven of those 12 contributions to our rich culture we inherited from the cultures close to water (especially the Mediterranean) and the trade routes from the East.

HAUSDOERFFER: *It sounds like, to return to the original question, you're saying you want to be an ancestor who returns humanity to a sunshine-based livelihood in relationship with how solar energy comes through soil, our plants, our livelihood, our connection; you want to be the ancestor that helps return us to sunlight and by "us" you mean Homo sapiens.*

JACKSON: Homo sapiens.

HAUSDOERFFER: *What are two or three things you want the reader to contemplate before themselves answering the question, "What kind of ancestor do you want to be?"*

JACKSON: Well, there are three old religious questions. First, what kind of a thing are we? Darwin took care of that for us. Secondly, where did we come from? The cosmologists have given us an answer to that question, in terms of the creativity from the time of the Big Bang. "What is to become of us?" is the third old religious question. That is the most important question for us now. Most of us will want our journey to continue on, but in an increasingly healthy and productive ecosphere. It seems worth knowing that we are the only species in this part of our sidereal universe that has become matter and energy's way to having gained self-recognition, and know about our origins in a scientifically verifiable way.

TRANSCRIBED BY CHRISTIAN AREL AUGUST 7, 2018
July 25, 2018, The Land Institute

POEM

Omoiyare
Frances H. Kakugawa

There will be no Nobel Prize for what we do,
No trip to Sweden, no medals, gold, silver or bronze.
But here we stand, past and present, preserving
For all generations, this lesson learned in what it means
To be human . . .
Once we abandon this heritage, all the years spent,
Day after day, year after year, in the shadow of the thief . . .
All would have been for naught. Bruised, frayed, tattered,
Like a flag after battle, we stand
With Human Kindness and Compassion,
A legacy for ages hence.

* 5 *

Earthly

Everything on the planet that is part of life and that sustains life is an ancestor.
VANDANA SHIVA

POEM

LEAF

Elizabeth Carothers Herron

You could mistake it for a moth, that laurel leaf
dried in a half-spiral
around the axis of its midrib, caught
on the fence and fluttering reddish
in October sun, easy to think *moth*,
the way they alight and stay, wings up,
quivering, though you know it's a leaf
flickering in the current of air
coming up from the creek.

What stays remembered later?
the muddy smell of water in the winter ditch,
the rough song of the fox at dusk,
the light on a leaf—pieces
alive in you that make you what you are.

Not your childhood or your illness or your degrees.
Not things known and visible, but this
moth-leaf you pause to watch
going out the gate—the quickness of it,
how very small it is, dry and half as wide
as the cat's ear, that same cinnamon
when light shines through. A leaf—
your life in it, fleeting and insignificant
and privately beautiful.

What is in your body rises
out of darkness and earth
to meet its fate. The leaf too
had a song, vibrating beyond the human ear
heard by bees, spiders and birds. There
on the fence it made something large,
something huge around itself.

ESSAY

The City Bleeds Out (Reflections on Lake Michigan)

Gavin Van Horn

> The best way to live
>   is to be like water....
> One who lives in accordance with nature
>   does not go against the way of things
> He moves in harmony with the present moment
>   always knowing the truth of just what to do
>
>   TAO TE CHING, verse 8

Gulls puncture the wind. On Chicago's edge, across this inland sea, I know Michigan rests somewhere beyond the line where chalky sky rubs against fresh water. The city loosens its grip here, bleeds out. Steel, granite, limestone, sand, shell, bone, tendril, hand—all will eventually surrender, become liquid.

The lake offers a pause. A curve of shoreline where the urban dweller can contemplate a humbling expanse beyond the buildings. When I come to the shore, I am overcome with gratitude that humans, so far as I know, haven't found a cost-effective way to build upon open water. The water says: no further. With that pause comes an opportunity.

The *Tao Te Ching* provides this counsel: be like water, it overcomes in the long run. "Nothing in this world is as soft and yielding as water / Yet for attacking the hard and strong none can triumph so easily / It is weak, yet none can equal it / It is soft, yet none can damage it / It is yielding, yet none can wear it away / Everyone knows that the soft overcomes the hard and the yielding triumphs over the rigid."[1] Ten thousand years from now? The city will crumble into Lake Michigan. It is the way of water. It rises, it falls, seduced and repulsed by a lunar affair, attentive to cyclical rhythms of which we are only dimly aware. People think you must go to the mountains to encounter wildness. There is nothing wilder than water. Even cities bow before water.

Released from the housing bust, construction has resumed again in the city. Downtown, anything close to the Chicago River serves as a hot spot for new buildings. Few distractions slow the momentum of a businessperson on a cell phone during a lunch break. But with the buzz of construction near the river, it's different now; people stop and stare. They congregate in small groups to look at cranes, soak in the sound of jackhammers, or watch the sparks fire from the hand of a welder. There is a dazzle to shining steel, muscled ironworks, and monuments of concentrated carbon. The motives driving the creation of these astonishments are simple. We network, drill down, and leverage for reasons similar to our ancestors': to cut the bite of the wind in halves and put dinner on the table. And so, we have pulled up soil, practiced chemotherapy on the land, squeezed the blood out of glacial stones, and sent the slag down the river to the Gulf, proud of words like *ingenuity, progress, scale, growth*. Other words are liquid whispers: *love, community, respect, care, obligation*. The water swirls on, above and below, brooding over the void.

The temperament of water is felt most directly on the restless edge of the city. I watch the lake's moods swing with the seasons. Storm surges releasing froth-violent waves that sweep cars from Lake Shore Drive. Ice floes that extend the shoreline half a mile, splintering aged wooden piers, carving out massive hunks of land. A quick shift of season and you will find surfaces so summer-smooth you'd think you could walk to Canada on a blue highway.

Water is a fitting symbol for Tao—without water we wither; it is the substance that seeks a bottom only to rise and recirculate in our every joint. We have saltwater in our blood, freshwater in our spine. The human body is about 60 percent water. Nearly 70 percent of the earth's surface is water. We are water tasting water, mystery tasting Mystery. Lake Michigan flows through my veins. It is the stream from the tap. I wash my dishes with it.

A city can take a toll on water. I ponder this when I find a beer bottle washed up on the shore. No message inside, apart from the emptiness of someone who mistook the lake for a dump site. But I also find fragments of sea glass, green and blue and brown shards shattered and pounded out and rubbed smooth as polished stone, the edge of our indifference made to shine like a gem. Water is as hard as it is forgiving.

Deep time awakens by a sea or lakeshore. Grains of sand and discarded shell homes, honed to a twinkling iridescence: one enters the wormhole of eons. The gulls play court jesters at this dance macabre, with thin laughing calls. Their generations have plied the shore long enough to be in on the open secret: all of us eventually go back to the water for polishing—with hopes of iridescence. "The world is nothing but the glory of Tao . . . / Rivers and streams are born of the ocean / All creation is born in the Tao / Just as all water flows back to the ocean / All creation flows back to be Tao."[2] Eons from now, what form will the feet take that pick up our fragments on new shores? Will they resemble feet?

*

Edges and cycles. On January 1, when calendars shuffle and attention shifts to resolutions and hangovers, I've taken up a new habit. I travel, not to other countries. I get out and walk, not always advisable or comfortable in the heart of a Chicago winter. But I take the arbitrary mark of a new year as a fulcrum and use it to get beyond pavement. Winter sausage and a pretzel roll, a daypack and a deserted train platform—a near-never happening—and I'm off, borne northward to a lakeshore refuge. Neighbors are throwing out Christmas trees, wondering quarterly worries about whether the economy will be beneficent in the coming year. At the lakefront, clarity arrives: there are time scales and then there are time scales. The Silurian, the Ordovician, the Triassic, the Quaternary, massive cyclical expansions and contractions of life. Ebb and flow. Dissolution and renewal.

We have to work to bend our minds around such durations. Familiar objects make unfamiliar scales comprehensible. Here's one way: If I throw my arms as wide as I can, try to embrace the wind pouring off the waves, and these outstretched arms represents the span of earth time, 4.5 billion years, then a nail clipper will suffice to cut human habitation off the body's map. Mind blowing, for the only species—so far as we know—with the minds to contemplate such metaphors of measurement. Sometimes I feel as though the water coughed up a grand epoch of mammals as an experiment, to see what we would carry forward into another cycle in the spiral of time.

To Lake Michigan I go to think bigger-than-me thoughts. To see my own limits transgressed. To imagine what water—the creatrix of land-born life—has to say, has to take, has to give. I come to dip my toe in the edge of where human control falters, fails, and falls into water.

On edges, feet crunch upon shattered shells strewn like a pulse before the water's persistent tongues. I dig bare toes into amniotic sand and grasp grace by kneeling, dirty kneed before a rising sun.

*

When I was in my twenties, I spent a year in northeastern Oklahoma on the Osage Indian reservation. The Osage people have a name for god that seems an honest assessment to me—Wakontah, one translation of which is *Great Mystery*. The Tao, Nameless Simplicity, numenon, Wakontah—they are all handles for a smooth door. "So deep, so pure, so still / It has been this way forever / You may ask, 'Whose child is it?'—/ but I cannot say / This child was here before the Great Ancestor."[3] The mystery child is incomprehensible but can be felt, touched, heard by the receptive heart and mind. If you take your net out, you will not catch it; you must be still and feel it tickle your ears and tousle your hair.

And seep between your toes. I walk on damp sand, marveling at what the lake returns to land. The spiral shells are the ones that astound me—the precision of the curvature, the love affair of genome and its expression in mineral homes, scrives a tally mark in favor of the Great Mystery. Surely a million other shapes would work, be as functional for survival. I tip my hat to the Mad Hatter of Life, Wakontah, who deemed the spiral a good shape to pitch out into the universe. A rococo touch from the strong hands of water.

*

I come to the shore when I need to breathe. When I need to have something breathe me. When I need a visceral reminder that the city doesn't have the last word. Water is patient.

The gulls are here again, my only visible animal companions. They circle with raucous confidence and wing me to America's western edge, where, half my lifetime ago, I kept company with their brethren whose eyes sidled my way. The open water and the gulls tug me between lives, between seas—the Pacific, Lake Michigan. Lake Michigan, the Pacific. I have a story about water.

Tides and a full moon—I was immediately drawn to her when we met. We were on the coast of Southern California. We were in college. She was self-possessed, had her bearings, knew injustice when she saw it, and her heart was giving. We danced around each other for a few weeks, met for coffee at a bookshop, went to a concert with a group of friends, then made plans for just the two of us.

We sat on the edge of America together one night, the only inhabitants on a shadowed beach, with the wind whipping through her dark brown hair, flying around her shoulders, her high cheekbones. Our feet were buried in sand warmer than the air. We clasped hands. I can see her thumb on top of mine, slowly drawing spirals. Her chin is to the wind. She is strong, mysterious, beautiful. I am falling in love.

Love is like water. We think it is like rock, but it's not. If we commit our life to another, we commit to water. We say vows as though pledging to rock, thinking that it won't move, that it can be built upon without worry. Water fissures this belief and asks, can you change with a person through their changings?

*

Can you expect answers from water? Many years later, in a desperate moment, having fled my gods long ago, I cast about for some portent, heaving a plea for guidance into the waves. I held the ocean responsible for bringing my truest love and me together. I needed the big water to keep its promise.

A gull pondered me, a quizzical look in his red-rimmed eyes, wondering if there were crackers in my backpack. I had none. He tacked left, then right, faced me again. I thought he might have a message to deliver from the ocean.

From ancient Greece to Tibet, India to Polynesia, birds are messengers, flying like dreams between our earth-bound existence and the realm of sky. *Augury*: the practice of looking to birds to know the will of the gods, from which comes the word *auspicious*. We take the auspices, grope for indications of benevolence.

Even within cosmologies that feature more anthropomorphic deities, the messenger birds persist—or parts of them do—in the guise of angels. Real birds, however, possess the advantage of being more visible than most angels, and therefore more consistently available for interpretation. And humans have interpreted everything, from scapulae to entrails to boiled chicken feet, on the lookout for hints about a prosperous course of action.

The gods' answers never come in plain talk, so some cultures developed highly specialized and elaborate systems, consulting the flight patterns, calls, and individual behaviors of birds, codifying these in formal charts and sacred texts.[4] In traditional Iban society in Borneo, for example, birds play critical roles in the lives of the people. Particular birds are believed to be gods (sons-in-law of the high god, Singalang Burong), and any activity for the Iban, from feasts to home building to warfare, is best accompanied by some assurance of avian favor.[5] Still, the Iban must interpret the signs correctly; they must strive to understand the language; they must treat the birds with respect and make appropriate offerings. And yet there's debate; in fact, "debating at length the relative merits of differing augural interpretations is an honored and favorite pastime."[6]

All of us, Iban or not, need assurance from somewhere, a favorable nod in our direction, an indication that we are on the right track. Think of augury as ethology with benefits.

Do you have a message for me? I ventured cautiously to the gull. *I don't need to fight a war or plant a field. I just want to know how to mend what feels as though it is breaking.*

The bird paced between me and the ocean, paused, offered a slight turn of his head. I waited. The ocean murmured. The gull remained silent. *Last chance*, I said, daring the bird.

Soon it became clear I'd get no magic. Gulls have their own business to attend to, their own company to keep. I wanted answers—from the gods, the earth, nature, our animal kin—some backing, some revelation from outside to quell the doubt within.

I instead received a different message: The world might indeed speak, but it doesn't speak to me alone. The gull is full of its own gullness. If that's not enough for me, I'm asking the wrong questions.

So, we looked at each other, the gull and I. We shared the beach and contemplated the briny smell of the wind together. The ocean murmured. I exhaled. *This is enough. Why ask for more when so much is given?*

*

Summer in Chicago. I cannot sleep. The light wakes me early. Mourning doves cry outside my window, plaintive *coo-ooos* that sound like a question.

I get up, ride my bike to the lake. I come to the beach to remember that feelings are older than humans, to have the water inhale and exhale me.

The water is smooth as a shard of sea glass. I sit in the shallows, water above my thighs, rocking gently with the pulse of the lake. Everything appears to be made of light. The reflective gleam of the rippling wind. The minnows that divide and merge in silver streaks around my knees. The rising sun that warms my back as I cry. Cry because I feel like one thing with the lake. The lake holds me. A mourning dove calls.

A shoreline is a place of constant change. Large waves are called breakers. The result of their work is in plain sight. A shoreline is a place where what is broken becomes beautiful. The work of water.

Ideals break. All ideals must. Our closest loves break. The expansion of love requires this. A shoreline is a place where what is broken becomes beautiful. Perhaps a loving relationship consists of equal partners, rubbed smooth and softened, like a pair of tadpoles in a circle; a yin and yang, each having the other buried in its core, incomplete alone.

We circle back—not circle, *spiral*, for we don't touch down in the same place. Each spiral we turn and ask the questions in new ways. We circulate from river to ocean to sky. We break and our edges soften, polished like sea glass. Two shards washed upon the shore, embracing each other. The work of water. I have collected what the edges of water offer: driftwood, sand, sea glass, the feathers and complaints of gulls, and the unremitting work of waves. "The best way to live / is to be like water / For water benefits all things and goes against none of them / It provides for all people / and even cleanses those places / a man is loath to go / In this way it is just like Tao."[7]

A large body of water is a great mystery. We cannot see too far under the surface. We cannot swim too long without a measure of danger. Where there is mystery, humility can flourish. If we think we fully know something or someone, we are likely to stop fully listening.

Part of the gift of water is that it reminds us of what we do not know. Perhaps this is why I come to the shore to listen. The waves susurrate a deeper knowledge, they sweep away my circle of thoughts, and sometimes, I think, I can hear the edges of the mystery that enfolds us.

I watch as the gulls dive and chase one another. On the shore, one throws her head back and squeals at the sky. The only other sound is the gentle lapping of the waves. I dip my finger in the shallow water and trace a spiral in the sand.

ESSAY

I Want the Earth to Know Me as a Friend

Enrique Salmón

I have many ancestors. Some were small-scale farmers. Some were freedom fighters against oppressive and genocidal Spanish Entradas. Some were teachers. I am referring to the Indigenous tradition of transferring cultural knowledge along the generations of my ancestors and to me through narrative. In my work as an ethnoecologist, I often reference this complex and sophisticated knowledge of plants, agriculture, and maintaining a relationship with the land, which is also a relative. I spent a great amount of time with some of my ancestors while growing up. They had a significant impact on my personality and the earth-centered paradigm with which I travel through life.

It is a paradoxical question to reason around the idea of what kind of ancestor I would want to be. Logic is inadequate when powering this notion and I can sense my multidimensional self touching upon some kind of minor enlightenment if, by chance, I were to reach a conclusion. First, before someone can become an ancestor one must have lived. How that someone had lived, I believe, is what is actually being addressed here. Living in order to become the kind of ancestor that one would hope to be is a work in progress. Similarly, life is a work in progress as we engage in mundane as well as life-altering experiences and share paths with people who remind us of things that we hold very dear.

A number of years ago I was part of a small group of Indigenous and other people visiting Arnhem land in the northern territories of Australia. We met up with two aboriginal guides who invited us to view a small cave holding 40,000-year-old rock art on its walls and ceiling. However, before we were able to see the rock art the elder of the two guides, Lefty, told us that we had to first be "introduced" to the land. The introduction took place in a nearby stream where we stood in very shallow waters that were gently pouring over exposed reddish bedrock. The banks were lined with local shrubs and shaded from the warm sun by eucalyptus and other medium-sized trees. Lefty dipped an old enameled cup into the cool water, sipped from the cup, and then proceeded to spray the water over us from his mouth.

In most situations in today's society I think that I might have been somewhat "offended" by this; a stranger spraying water all over my face and body with water that they had just been holding in their mouth. But in this case I, as a Native person, recognized that we were in ritual space. In this space meaning and symbols change. Spatial-temporal portals into alternate dimensions of reality open, and most importantly, one's awareness is expanded. In ritual space, one can more easily understand how all things are interconnected and water being spit onto you loses its gross factor significance.

Lefty had explained beforehand that the bees of that landscape must first get to know us before we could proceed onto a sacred place. The bees, the water, the plants, and the land are all part of a dynamic multilayered cosmos operating under a process known only to itself and to those observant enough to recognize it. The bees, and therefore the land, had to get to know us first in order for us to proceed into their space. For the uninitiated: think of a situation such as the first time going alone to a restaurant in a foreign country. A lot of things such as tables, chairs, the wait staff, drinking and eating utensils seem the same, but the customs are different. You will feel uncomfortable with certain ingredients in many of the dishes. You might not know how to get the attention of the staff and even how much to tip the waiter. Perhaps you cannot understand the menu and there might not even be a menu. As a result, the staff and the local customers will not be sure how to react to you. However, let's say you return to the same restaurant, but this time with a local who knows everyone who works there and also knows many of the customers. The person is able to guide you with local restaurant customs, ingredients, and how to treat the staff. Over time, you will be a regular and treated as a welcomed guest to the restaurant. Like Lefty's bees, the restaurant and its locals need to be adequately introduced.

I rather liked the concept of being introduced to the land in Arnhem, immediately understood it, and was reminded of something I wrote in an article entitled, "Kincentric Ecology: Indigenous Perceptions of the Human-Nature Relationship."[1] In the article I drew ideas and concepts from my own tribal worldview of our relationship and responsibility to place in an effort to explain how American Indians see themselves as part of an extended ecological family. I focused on the Rarámuri concept of *Iwí gara*. I wrote that, "Iwí gara then channels the idea that all life, spiritual and physical, is interconnected in a continual cycle. Iwí is the prefix to iwí gara. Iwí gara expresses the belief that all life shares the same breath. We are all related to, and play a role in, the complexity of life."[2]

The notion of iwí gara implies that humans are not the only facet of the natural world that has an important part in the complexity of life. This concept is a base metaphor that influences how I lead my life and guides so many of my decisions regarding work, getting to work, shopping for food, whether or not to grow a garden this year, shopping at the farmers market, buying clothes, which coffee to drink, and who to vote for in local and national elections. Knowing that I am related to everything around me and share its breath leads me to focus on my responsibility to consider all living and nonliving things with regard to choices and actions that I might take. I am, therefore, one of many stewards to the land and natural worlds that I share breath with and minister to in appropriate ritual, thought, and ceremony. In this way I feel that I am a champion, a kindred spirit, a defender to, and even a compadre and carnal with the Earth, and hopefully, the Earth returns the feelings.

The Latino concepts of compadre and carnal were included intentionally. Both activate culturally sanctioned mental spaces related to concepts of friendship and companion. However, buddies who are compadres and carnales are more than "pals" and "buddies." They are terms of endearment, spoken from the heart. To be a carnal implies responsibility and commitment to the friendship. When I announce to a group, for example, that a particular person, "es mi carnal," I am telling everyone that this person is more than a best friend. I am also saying that I know that this individual will always be committed to our friendship no matter the situation, no matter how awful a person I might be behaving as, no matter the costs, and that I will always return the same responsibility in kind. I am also saying that I know that I can always count on this person to be my defender and champion, and vice versa. To employ an old rock climbing metaphor, if I had to choose someone to be at the other end of my rope it would be mi carnal, because I would feel secure knowing that they would do whatever it took to not let me fall. I would hope that the Earth thinks about me in these ways; I want to be one of many of Earth's carnales.

I suspect that people who are drawn to certain places and landscapes and yearn to get to know them spend lots of time and energy learning the plants of a place, the geology, its unique weather patterns, and the animals endemic to that place in an attempt to feel connected to and a part of the place. However, I wonder how many people stop to wonder what that landscape thinks about them. We may come to "know" a place, but does it know who we are? Is the cycle of Iwí completed when the land knows us as a friend, a carnal, and maybe even a relative? This is what Lefty was trying to open up to us when he ritually introduced us to the local bees.

I want to be the kind of ancestor who is known by the Earth as a friend, a relative, a carnal. I want us to be the dearest best buddies; always watching each others' back. I want the Earth to hear and desire the songs that I sing to it. I want it to appreciate my garden of ancestral foods that I grow in a manner that returns nutrients to the land. I want the Earth to recognize my friendship through the kind of car I drive, my visits to the local farmers' market, when I recycle things, by the kind of toothpaste I use, and when I take time to teach others how to develop a relationship with a landscape and to behave as a good, caring relative. I believe that the Earth is a living being. People can come to know its personality by spending time around certain landmarks and places that are encoded with its power (spirit). Some of us who are trained and initiated act to minister and steward these places of power. The rest of us can act to maintain and encourage the resilience of our favorite places; we can be their friend. If the Earth is alive and conscious it will recognize who its cohort of friends are.

I have read some non-native ecological, environmental researchers and metaphysical writers who agree with the general indigenous notion that the Earth is a living being. However, too often these same thinkers treat these concepts as if they are merely linguistic cultural metaphors. They suggest that the ideas are not based in any tangible reality, but are only abstract and intangible beliefs. I believe that the kind of people I just referenced mean well and have the best intentions. Unfortunately, I wonder if their Western thinking gets in their way. Western scientific language, and therefore thought, is deterministic leading to linear thinking. As a result, one would have to search for very long among Western constructed categories and domains of the natural world that are not linear. This is important because it has been my experience as an Indigenous person, and among other Indigenous people that I have worked with, that the land, places, and realms of power and the concepts that surround these ontologies are very tangible—often existing alongside us in both this dimension as well as in parallel spatial-temporal dimensions.

The Earth is a multidimensional being and, therefore, affords multidimensionality of knowledge in order to understand it and to feel connected to it. What this means is that friends of the Earth should approach their relationship with it along various and divergent avenues. One can begin this journey first by simply sitting with a place over and over again. I am a college professor who teaches American Indian studies classes. One such class is named American Indian Science. American Indian Science approaches an understanding of the natural world along multidimensional paths of observation, one's self-identity that is connected to a place, spirituality, and shared experience. I had been struggling with how to help my students gain, at least, a small taste of experiencing a Native understating of place. This is difficult to do in an urban environment where I teach and where most of my students come from.

I designed a course-long assignment which asks each student to identify a location anywhere near where they live or somewhere on campus and to view the sunrise or sunset. Each week they are to sit or stand at that spot. They are also supposed to be at their location at the same time and watch the sunrise or sunset. At the same time they are asked to make sure that their location has a clear view of the western or eastern horizon that includes a tree, telephone post, or geologic feature that they can use as a reference point. I also ask them to be aware of all of their senses, allowing them to stretch out to everything that is occurring around their spot. Some students choose to look out their back doors. Others find a location around their neighborhood. One student chose a parking area on the eastern end of a bridge that crosses the San Francisco Bay near my campus. Others choose to climb one of the hills on the south end of campus in order to get a clear view of the bay and of the western horizon.

While the students are engaged in this assignment they write a weekly journal detailing their experiences and thoughts. I never lead the students toward what they should be experiencing. Their only direction is a short set of questions that they should respond to each week. I ask them to note what they have seen, heard, or felt. I ask them to notice any changes. I also ask them to connect their weekly experience to things that we have been reading about or discussing in class. While reading their journals I find that most students are a bit confused at first; wondering why their professor is asking them to engage in this exercise and what it has to do with "Indians." However, after a couple of weeks I have noticed shifts in the students' reactions. They begin to make note of subtle changes in their space. They begin to include comments about how they are "feeling." Some note that they find the exercise relaxing and a nice escape. One student noticed a deer approaching a bush each week and finally decided to research what kind of bush it was and why deer like to eat it. One student found that each time she sat at her spot on a hill near an open space, the same elderly couple would stroll by holding hands. One week the couple did not show and the student noted her concern for them. By the end of the course most students reported that they wished to continue the practice as they found it a welcomed respite from their daily urban and scholastic rigors.

What my students are reflecting is a small example of multidimensional experience with the Earth. To know a place and to become connected to it as a friend requires more than knowing and recognizing its parts. It requires an openness to the messages and knowledge the Earth is transmitting.

No matter the realities, intangible or tangible, spatial and temporal, the Earth will house all of our experiences and thoughts because it is the source of all of our experiences and thoughts. And if the Earth is the source of so many of our experiences, values, thoughts, and even our identities, then I recognize that it is central to my lineage; it is an ancestor. There is no better example of the kind of ancestor I would want to be. Somewhere in that library of the Earth's memories I hope that, once I leave this realm, my ancestry is placed in a file entitled "relatives, dear friends, and liked by bees."

ESSAY

The Apple Tree

Peter Forbes

"That tree is everywhere; I love how it lives on in our community."

My neighbor speaks of a 150-year-old apple tree that hundreds of people have picnicked beside, countless children climbed, at least two people got married under, and which produced hundreds of gallons of cider until an icestorm split her in two and brought her crashing to the ground. Then I spent three years slowly making 120 thick presentation boards and dozens of spoons and small pieces of furniture from her. My wife's and daughter's necklaces hang from hooks I made of her wood, all of my hats hang from her beautiful wood, my tools now have handles made from her. Though the stickered pile of her lumber is now completely gone, people still stop by our farm and ask if there's any of her left. "I have a niece who's getting married next month, could you make one last . . ."

It must have been one of the McLaughlin women who planted that Tolman Sweet around 1850, a practical and hopeful way forward after a life-and-death voyage fleeing the Great Hunger of Ireland. I imagine her with two buckets of water hauled in balance by a wooden yoke watering this young apple tree while her husband and son were fighting in the Civil War. It would be hot, the grass deep, and the walk from the spring slow and sweaty, but she did it every day, making peace with the torn world around her, unaware of what the future would bring, but still doing what her instincts told her to do.

I want to be the sort of ancestor that plants apple trees for somebody else. I want to create beauty and health on the land that someone may recognize 150 years from now and be inspired, no matter what's happening in their world, to do the same. I hope to leave behind a key to true wealth: a set of relationships, knowledge, and access to places that enable someone I don't know to make a good, fair life.

I wonder what it means, today, to be knowledgeable, and what is the raw material of our knowledge? We seem to believe we know it all right now, that we are experts, but even in our certainty we destroy so much. I believe that we are in a never-ending process of slowly becoming knowledgeable through direct human experience. But what, today, do most of us have direct experience of? What unlearning must happen for us to experience it?

I imagine that a "modern era" begins when subsequent generations look back on the one before with some version of the question, "How could we have possibly lived that way?" How could we have openly, legally created an economy based upon human enslavement? More recently, what were the conditions that allowed more than 4,000 men and women of color to be lynched in public spectacles in our country? How are we perpetuating those conditions today? What are we doing now that will be judged by our descendants as morally bankrupt?

The wisdom to confront these questions arises from a courage to see myself from a different perspective, which I get from nature and from people. Cultures, whether we label them good or bad, are never trivial because they are ways of seeing the world. When I see myself through a different lens, I may not like what I see but I'm liberated to be something better. It's from my own working relationship with nature that I've made mistakes and reconciled them, made a huge effort and been cared for, learned respect by seeing how much bigger the world is than me. My heart has been touched by beauty; my head reached by the power of cultural and natural diversity; both have deepened my sense of humility, belonging, and fairness. This knowledge of what I don't know has made it more possible for me to stand beside people who are different than me.

It is my direct experience of nature that has helped me to climb out of the blindness of my own isolated culture and opened me to the claims of others; I credit my contact with difference in nature and people as what has enabled me to evolve and grow.

The soul of our country is born from epic choices about our relationship to land and to one another. The relationship is good when it's about respect, care, and equity. It's bad when it shows us stealing from our children for ourselves, and it's ugly when it alienates anyone from the genius of place on their own terms. To seek a healthy relationship with the land through how one lives, how one eats, and whom we welcome to the table or not, is transformational because it ultimately *is* about love and healing. It's about relationship. And most people get this because we humans—at our core—are more tuned to relationship than to isolation.

What does it mean for a life to go well? What does it mean to have a life of significance . . . a life well lived? The deeper my relationships with other human beings and with physical places, the fuller have become my answers to those questions. This has never been easy and the questions keep getting harder and more personal. The toughest ones I don't have answers for today. What aspects of settler culture, to which I'm a part, need to change and evolve for everyone's well-being? How do I come to be welcome on land that I have stewarded well for decades but was stolen from someone else? I know the name of the European immigrant who first built a cabin on this land in 1804 but I don't know the name of the Abenaki mother, father, and daughter who were forcefully displaced from the land they also loved. With this knowledge, how do I bring my gifts and meet my responsibilities? Wrestling with these questions because they matter is not about guilt or shame, it's about my journey to become a whole human being through knowing history, understanding my privilege, and using that privilege to not repeat history.

While I'm a grateful product of institutions like National Outdoor Leadership School, places that introduced me to wild beauty and nature, I no longer believe that the world needs me to "recreate" myself in nature; the world needs me to get to work. Today, the world does not need me to "leave no trace"; but to leave a beautiful trace: the demonstration that I can create beauty and lasting things though my relationships with nature and people. Let my trace also be my voice, expressing why it's not ethical or practical to love nature but to disrespect people, how doing either alone leads to the destruction of both.

I want my descendants and yours to have both cultivated and wild places where they are welcome to work, to hunt, to harvest, to grow, to share, to learn a relationship with nature where they are not consumerists (recreating themselves) but producing things that are real and beautiful. When I consider our world's challenges of disconnection and alienation, what our culture needs—as much as democratic institutions like our national parks and public open spaces—is a renewed belief in long-term familial relationship to place; which is to say direct human relationship to the land and waters of our planet through the generations. We cannot rely on institutions to protect land and nature; we must learn how to do that ourselves.

You can't take anything from a national park but memories, because we don't trust ourselves anymore. We abdicate our fundamental responsibilities to a federal agency or a nonprofit corporation when we personally need to develop the skill for how to dwell on this earth, to see ourselves, always-already, as ancestors. To go into the forest or into the ocean and to bring out wood, protein, or energy for your family leads to love and fierce care of that place. Yes, of course, it's true that we humans become greedy too often and turn nature into things to be bought and sold and that ideology—capitalism—can hurt us and nature badly. Run amok, capitalism contributes to another ideology called conservation, which says you can destroy that over there as long as we save this over here. And that separation of deed and creed leads to the easy temptation of living largely outside an experience of nature, but believing you are respecting nature by sending an annual check to your favorite, well-branded, environmental group. How do we ask more of ourselves?

When I know how to do something practical, I'm more able to share myself. Whether you live in a megacity or a village, making and doing is a type of currency that when given freely enables us to participate in a world of relations, human and more than human. Helen, my partner, made this clear to me last night when she told me how she has learned over time that the very beautiful swallowtail caterpillar with its breathtaking orange dots loves to eat our parsley so she moves these caterpillars out of our garden into the field to eat wild carrots and Queen Anne's lace, which is in the same family. It's knowledge that enables life, ours and theirs, to grow. I was fortunate enough to go to high school and college and those places taught me very important things about how to be a writer and a photographer and a student, but mostly they positioned me, not prepared me, for life. The wild and cultivated places in my life have taught me the most, they have prepared me to live my own unique life, and they've given me useful knowledge to share with others. There are days when I doubt the scale of my service and wonder if am I being nostalgic—what difference can this apple wood make in a world of bits and bytes? And then fellow travelers on the front lines of change come wearily to our place to renew themselves, to heal and to re-member, and I forget my doubt.

What will she look like, that great-great-great-granddaughter of mine? She will be browner than me, and stronger than me. I have no fears for her other than the fears I have for myself: sliding into the disconnection and alienation of any culture that consumes more than it produces and protects. I do my days in a certain way so that my descendants will know place-making: how to make soils healthier and more productive, how to grow enough food for themselves and others, how to make art and beauty from our world, how and why to share the complex story of a life lived in place, how and why to welcome strangers. I want my great-great-granddaughter to know how to do practical things for her own benefit but also to earn and to convey respect. I hope she carries the deep intuition that her presence creates beauty in the world, that she has the skills, the humility, the knowledge, and the relationships to do that.

All beings take different trails, and that's good. I want to leave on the land, and in my descendants' hearts, a trail to follow that helps them to find themselves. I want my life to be a reminder of where to find those trails when they are needed, especially when hidden by the world we have created, when the more-than-human world is less and less. These are the trails to follow to remain connected to ourselves and to life itself. My intentions and wealth are in the placement of stones, the willingness to plant apple trees for the future, to love land and people well, to share it all, and to learn how to return it to those not yet born.

ESSAY

Humus

Catriona Sandilands

Behind the little house that I rent on Galiano Island, there is a back trail cut through the woods from the road, up a steep hill, to The Bluffs. The Bluffs were set aside by the Galiano community in 1948: they were one of the first places in the Southern Gulf Islands to be formally protected, by settler-colonists at least, from logging and real estate development. You can see why as soon as you arrive at the lookout: The southwest view across the Salish Sea to Mayne, Pender, Prevost, Salt Spring, and Vancouver Islands—and, then, on a clear day, into the San Juans and the Olympic Mountains—is deep and seemingly limitless.

On the way up to The Bluffs, from the house, the back trail winds through a dense, mostly second-growth fir and cedar forest, across private property marked with polite signs prohibiting mountain bikes, back and forth up several hills and plateaus, to the top. It is called the Grace Trail. The trail is named, rather controversially, after one of the people who owned the property in the 1990s and who wanted to turn the place into a residential community; the trail access is a direct result of their development aspirations. The plan failed for complicated reasons (Galiano is full of complicated reasons), but the name, Grace, remains. I like to think of the grace of the trail a bit differently, though: whatever you think about the origins of the trail or its name, it is a generous ascent through a landscape that is not in the least bit pristine but that is so full of life that it still feels like a sort of biodiverse blessing at every step.

There is a particular spot on the Grace Trail, about three quarters of the way up, in which I always stop. I can't help it. On this small plateau, there is a grove of older cedars and firs, their lower limbs dead and dying, that is roughly encircled by fallen cut trees, and then circled again by erratics dropped there by glaciers. In this grove, there are the trees in various stages of decay, and there are also the finely filigreed mosses and lichens—beaked moss, ragged lichen, beard moss—that thrive on the rotting wood. And then there is the forest floor, full of slowly rotting vegetation. There is so much matter, and so much time involved in its turning-to-soil, that the ground is perpetually spongy: branches, cones, bark, bryophytes, a million tiny invertebrates, and a hundred times that many invisible microorganisms, form a bed of always-in-process decaying that is not hard to describe as a carpet, it's that thick, that soft. The place is an absorbing bowl of silence as a result: generally, the only sounds that you hear are the passings of the Victoria to Vancouver shuttle planes. In the quiet, the smells become audible: a high soprano of wet cedars, a funky tenor-baritone harmony of Douglas firs and saprophytes, and, in the middle, a subtle alto of fallen needles.

In this place, decomposition is especially articulate. I am no soil scientist, but even with my limited understanding I can see, and smell, that this compost is not just random rotting, some unorganized passing of one generation of lives, through death, into the next. In a complex, self-organizing calculus that depends as much on atmosphere as internal composition, some organic matter is quickly and biologically reduced to minerals for vegetal uptake: nitrogen, carbon, phosphorous. A small amount of matter, however, remains in the ground as relatively stable humus; very stable humus can hang around for hundreds of years, its chemical wealth safely locked up in organic chains, relatively impervious to erosion. Mineralization is essential for immediate generations of new life: you can't have a forest if there aren't nutritively rotting trees. But the generosity of humus is more complex: it is crucial to soil structure and condition, moisture retention, nutrient storage, and even temperature, but largely because it does *not* give up its mineral wealth quite so readily.

Donna Haraway has, rather famously, declared that she is "a compost-ist, not a posthuman-ist: we are all compost, not posthuman."[1] In contrast to a view that, once again, manages to place the human at the conceptual center— humanities, posthumanism, Anthropocene—she understands compost as a metaphor for, and also a medium of, conceptual and corporeal mingling, in which many kinds of lives and deaths can come together to germinate new, multispecies possibilities for kinship. (Indeed, the composted matter already on the forest floor is an ongoing gift from our mingling ancestors.) Being good kin, being good compost for others especially at this critical moment, requires mourning; as Haraway writes simply, "There are so many losses already, and there will be many more,"[2] and just turning our attention to loss is an important, if painful, act of ecological understanding.[3] But mourning is not a just a matter of acknowledging death and moving on to life enhancement again. Rather, being involved in relationships of care with and for multiple others involves being deeply aware, on an ongoing basis, of fragility and precarious interdependence as the condition of all of our lives, and deaths, together. To respond, we need to take on an attitude of what my friend Louise Squire calls *ecological death-facing*: "To face death is thus, paradoxically, not to die but to view life differently. By switching from a mode of destruction to one of acknowledging the reality of the world and one's own finitude, life emerges as a reconstructive project."[4] In this way, composting—a death-facing practice if ever there were one—suggests possibilities for developing a compassionate appreciation of the ways in which we are ancestors to the world (and the world ancestor to us), not just of our own kind.

When I began to write this essay, I asked Louise what she thought about compost as a figure of ongoing ecological ancestry. With her particular Buddhist thoughtfulness, she suggested that the mere recognition of our own death, decay, and potentially nutritive corporeality for others is not sufficient for a truly attentive response to either everyday or epochal loss. She told me about an image she had seen of a flower growing out of a mass human grave, and that she had hesitated at its too-easy indication of hope springing out of horror. In order to live and die well, she argued, we need to respond consciously and with integrity to the ecological and social devastation that is the present, as well as to the inevitable end of our own lives and relationships, so that we can be *good* ancestors: ones who thoughtfully cherish and nurture the world and the liveliness of others as best we are able, and whose gifts in the present may resonate in new ways in the future. Directly opposed to the prevalent, heteronormative ethos of *inheritance* in which, among other things, we are led to believe that we are justified in using the world instrumentally for the direct, accumulating benefit of *our own* future generations (part of what queer theorist Lee Edelman has termed *reproductive futurism*[5]), this view underscores the impermanence of any and all of my direct contributions to the world even as it also demands that I work consciously toward relations of compassion and justice in the present. Inheritance is possessive individualism passed on down the patriarchal family line; good ancestry is, in contrast, a generous offering of one's life to an unknown, multiplicitous future. You can't take it with you, and you can't even really pass it on. But you can help to create a medium in which lives may flourish in the distant future by living a life well, and with an eye to its eventual decomposition. What will be the quality of the soil that I cannot help but become? What will my present mean to an inevitably, and increasingly, unpredictable future? What does it mean to think of myself as an ancestor among ancestors, both the recipient and the source of nutrients?

Perhaps because much of my life has been lived in queer forms of community, I have deep reservations about the language of "saving the earth for one's children." I am grateful for the many gifts that my parents have given me; I also love my daughter and work hard to help create, with her, conditions for her flourishing (sometimes emulating my parents, sometimes not). But I have seen the damage wrought by zealous attachment to family and heredity; I have felt exclusions and violences, both petty and profound, made in the name of protecting what literary critic Lee Edelman terms The Child, an innocent to be isolated from the world, rather than immersed in the messiness of it; and I have also had the privilege to witness, and be part of, the rich communities of care that have sprung up, for many people, in the wake of sometimes traumatic alienation from families of origin. The practice of nonfamilial kinship is by no means confined to LGBTQ relationships (especially given the current focus in much mainstream North American LGBT politics on marriage and parenting); there is, however, as J. Jack Halberstam has emphasized, a particularly rich archive of queer relationships and ways of life on which to draw in order to imagine, and practice, a more expansive understanding of kinship and ancestry.[6] For me, as for Haraway, the only viable understanding of kinship out there, in opposition to a human-centered Anthropocene, is also one that includes thoughtful attention to the flourishing of other kinds of life. It is not merely that my child's future depends on the wise use of resources; it is, rather, that an orientation to others in which I may also be a resource for them, in networks of reciprocity and relationship, is key to our collective survival. Of course, this includes what I leave behind.

Although the figure of compost certainly captures some of this aspiration toward multispecies future-making, *humus* turns out to be a better metaphor for/medium of my vision of good ancestry.[7] To me, it speaks to a sort of *future-oriented* decompositional practice. On the one hand, there are my thoughtful actions in the present, oriented to multispecies flourishing, to new entanglements of kinship, to challenging the reproductive futurism that, as I have argued elsewhere, links capitalist accumulation with specific ideas of human generation in ways that are horrifically bad for the planet.[8] On the other hand, any lasting humusian infrastructure to which I might contribute depends on the transformation of my work, along with that of others, into *new* organic forms that are not echoes of the old, but rather the results of the work of future communities of macro- and micro-organisms, who will take from my life and work not only the immediate chemistry of my body as it passes into dust but also, I hope, a more complex infrastructure. Drought-tolerance. Density. Structure. Maybe some leavening lyrics to aerate the soil of environmental justice work that cultivates flourishing, complex communities. Maybe some slow prose to absorb the quick acid of the internet. Until the humus, too, erodes and dissolves to minerals.

Although I love The Bluffs, the long, penetrating view from the top is not the right place from which to reflect on the kind of future I want to help generate. Clarity of vision is inspiring when it breaks, through the fog of the everyday, into my thinking, and I am grateful for it. But it's the wrong perspective for a careful reflection on my possible contributions to a flourishing, future, multispecies world. The right place for ancestral thinking, for me, is the ragged circle in the forest on the Grace Trail, where I can smell—and even hear—the diverse ancestors who have come before me, from whom this place takes sustenance, and in the midst of whose eloquent decomposition I feel compelled, on every ascent, to pause.

ESSAY

Building Good Soil

Robin Wall Kimmerer

She always listens, my elder. I go to her with so many questions, my *Nokomis anenemik*, my Grandmother Maple. I'm a grandmother too, feeling the weight of a future in which the patterns of the natural world can no longer be relied upon, when so much we love is threatened, when I fear for my grandchildren and the grandchildren of warblers, bumblebees, and hemlocks. It is a comfort to consult with intelligences other than our own. She has seen more than two hundred sugar seasons; her bark is furrowed deeply and torn away on one side where the wind took a huge branch a few years back, revealing the dark crumbly core of her future. But, she is still here, steadfast in doing what she was meant to do. A nightshade vine grows from that nurturing well, red berries gleaming in the sun. Herself the ancestor of thousands, who better to give counsel in these urgent times, on the edge of climate chaos and the age of the sixth extinction? What does it mean to be a good ancestor?

I settled into her leafy lap. She put her arm around me and gave me her answer. I saw it in the medicines, thick and plentiful beneath her canopy, in the cushions of moss that upholster her branches and the red-backed salamander wiggling in the hollow of her roots. I read it in the crowds of maple seedlings rising around her. "To be a good ancestor, you have to build good soil."

Look at this ragged and resistant old maple, hit by lightening, heart rot, and time and you know one true thing—that the world is constantly in flux. We cannot know what lies ahead, especially as we hurtle toward a catastrophe of our own making. So what matters most is the potential for regeneration. Building good soil enables resilience in the face of change, buffering against shortage and stress, so that life force can go into something more than survival . . .

into becoming. It is the work of ancestors and ancestors-to-be, to support the becoming of what they cannot imagine, but trust will arise. Building good soil means preserving room for possibility, for a world open to creation again and again.

I remember that after the blast at Mount St. Helens, after the searing wind when all was laid waste, that life began again, in tiny pockets of good soil. I remember that when the fires reduced forests to ashes, the seedlings arose, in pockets of good soil. I remember industrial waste sites and mine tailings where the soil was obliterated and the community of life has never recovered. I know her words are true. To my maple grandmother, building good soil means protecting the sources of regeneration.

She should know because building good soil is what maples do. They are well-known for the moist fertility they create where their leaves come to rest. Imagine the tons of leaves that each fall blanket the forest floor in yellow, orange, and scarlet. Before leaves let go from the twig and swirl to the ground, they have a choice of what to do with all the summer's goodness they have accumulated—use it or lose it. The leaves can re-translocate or send their nutrients and sugars back into the parent tree to be used again next year or they can keep them in the leaf and let it fall to the ground. Trees like oaks and beeches are a little stingy. They retranslocate nutrients, so the dry brown leaves they drop are just skeletons; they rattle like dry bones on the forest floor and take a long time to decompose because they're simply not very tasty. But maples make a different choice; they keep nothing back for themselves and drop leaves, which are a feast for soil citizens, who turn them quickly to humus. Which feeds beetles and salamanders and trillium—and baby maples. I want to be that kind of ancestor, giving it all away to the future.

Soil, to me, is a worthy ancestor for it is simultaneously the repository of what has come before and the garden for what is yet to come. It joins the realm of memory to the ultimate source of becoming.

As a grandmother, an ancestor-in-training, I ask myself, "What do I love too much to lose?" and "What will I do to carry this safely into the future?" In becoming ancestors, we honor our obligation to the ones who will follow, the ones who will carry our story into the future. We build good soil so that they can continue. For them, I want there to always be thrushes and strawberries and glaciers. Perhaps it is a kind of immortality project, to keep the web of relationships intact. But as an ancestor, I feel a responsibility not only to descendants, but to the ones I'll call ascendants, the ones who will rise above us, the ones not yet dreamed into existence—they who will surpass our ways of being because the earth has changed. Maybe they'll build good soil out of plastic or blossom in the ozone. I don't know what they will be like, what hardships will shape them, what beauty will lift them up, what lessons they will learn. But I know one thing our ascendants will need, in order to ascend. Whatever the nature of their world, they will need to know how to love it. Better than we have loved ours.

Grandmother Maple, my teacher, eventually gravity and fungi and one last tap by a woodpecker will bring you down, then you'll be firewood, warming this grandmother, rocking her grandchild in front of the woodstove and telling him of his ancestors whose stories lie just below our feet, remembering so they can nourish us still.

The regenerative capacity of good soil lies not only in its organic matter, its nutrients, its microbes, sand, silt, and clay. Good soils hold a seed bank. They hold ideas. A seed bank is what ecologists call the reservoir of dormant seeds, spores, and other propagules that have accumulated in the earth over time. They are the seeds of windblown grasses come to rest, the pits pooped out by passing birds, the rain of seeds from nearby trees and spores of faraway mosses. They are a history of what has grown here before and the future of what might grow here again, depending. Depending on the soil we build and the seeds we save.

Biodiversity is the imagination of the earth; it propels the emergence, the evolution of our ascendants. Building good soil means preserving a world rich enough in biodiversity that it can imagine itself anew.

The happy truth is that when I am an ancestor, I will be soil. Human become humus. I view that as a wonderful outcome, to mingle with roots and translucent springtails, become entangled in mycorrhizal networks and commune with bits of ancient mountains. This is company I could relish for eternity!

It will come soon enough. Meanwhile, on my way to becoming an ancestor, I want to seed the soil with stories of love for the earth, for it is our failure to love the Earth enough that allows us to let it slip away. Following the lead of Grandmother Maple, I want to lay myself down to build good soil, so that that all the young ones yet to come, salamanders and maple seeds and grateful humans, have a chance of home.

INTERVIEW

Vandana Shiva

John Hausdoerffer

HAUSDOERFFER: *I'd like to just ask you the title of the book: what kind of ancestor do you want to be?*

SHIVA: I want to be the kind of ancestor who has future generations to be an ancestor to. Because that's the big issue now. And that's why a lot of my life's work is to ensure we have some possibility, some potential of a future. Because on the path that we are walking on we are absolutely making sure that maybe two generations from now there isn't a generation for any human ancestor. So I'd like to be an ancestor.

HAUSDOERFFER: *So just being an ancestor, by having a humanity around after you are gone, would make you a good ancestor?*

SHIVA: Right now. Right now that is the situation. Maybe years ago if you had asked me this question I'd say I'd like to be the kind of ancestor to human beings that allows other beings to flourish on the planet. I still say that. I still do it. But I think we've pushed so many species to extinction. And we're taught we're privileged. You know the superiority, the anthropocentric thinking. That we are above other species. Just like there's now a return of the colonial, white supremacist thinking. That we are above Black people, Brown people, the Spanish-speaking. The problem with that kind of thinking, which is very, very much the thinking that someone like Hitler had, where he thought, "Well, get rid of so and so, get rid of so and so, get rid of so and so because they are inferior to me." That illusion of superiority over other species and over other humans lends itself to a sense of disposability of the others because you don't understand your interconnectedness with others. You don't understand your mutual interdependence with others and so you basically say, "I can remove this, I can remove this, I can get rid of the butterflies, I can get rid of trees, I can get rid of the ocean's health, and I'll thrive, or I can get rid of other human beings and the children of other people, and I'll thrive." And you don't because your thriving depends on others.

In yoga, when you breathe in and breathe out your breathing in and breathing out goes with *So Hum*. *So Hum* means, "You are, therefore I am." You the tree are, therefore you give me oxygen; therefore, I can breathe and am alive. Otherwise, I would not exist by being an emitter of carbon dioxide on my own. My biological being is an ecological being, an interbeing, and it includes the tress and the plants. Just because of the ability to use huge amounts of fossil fuels, the fossil mind has created the illusion of separation and mastery. No, I am not, separate from the tree. Without a tree and plants, and their power of photosynthesis, there is no process of the absorption of the carbon dioxide we emit. So right now I'm at the point where I realize this process of disposability of the other has reached such a climax that "the other" has become our own future generations.

HAUSDOERFFER: *It sounds like you're saying that the ancestor that you want to be is one that has an interdependent humility. You have the humility of learning from other cultures, of connecting with the intrinsic value of other species. It reminds me of an ironic consequence of anthropocentrism. I was just in Sunder Khal, India, outside of Majkhali, and the British had historically pushed the monkeys off the plains and into the mountains, and now the monkeys are so populated that they're destroying the crops and the farmers of Sunder Khal are now starting to have to move off their land. It's almost like anthropocentrism, pushing monkeys off the plains for human development, is now leading to human displacement in the mountains.*

SHIVA: The real problem is, when you don't understand life is in cycles and when you don't understand that life is in webs, you break the cycle and you cut the links of the web because of your illusionary arrogance of being at the top of a pyramid of power. You will eventually be acting against the human interest. So it's stupid. Anthropocentrism by definition has to be stupid and blind.

HAUSDOERFFER: *The actual question, what kind of ancestor do you want to be, was first posed to me by Winona LaDuke's neighbor, Michael Dahl, on the*

White Earth Reservation. What I've since learned was that this question comes from the Seventh Generation ethic of the Anishinaabe. Do similar kinds of questions emerge from spiritual practices in your region, whether they're Hindu, or Buddhist, or Bihari?

SHIVA: The Seventh Generation is built into Indian civilization. Exactly the same concept. When you take any step, tread in a way that you do not leave a negative impact for seven generations to come. So the Seventh Generation test is very, very deep in any society that has lasted over time. Sadly, we're all been imposed with this nonsensical idea of development. Every destruction is justified in the name of development. To create the debt trap, they tell us we are lacking, then they send in plastics and pesticides. Now everyone is running to get rid of plastics, from the oceans, from the towns. Mumbai has just banned plastics and the whole city is going crazy because when you build an economy around plastics, you don't make a one-day decision to ban plastic. It doesn't work. So, the Seventh Generation logic teaches us that you could be an Indian in the original land of India or you could be an Indian in the land where Columbus landed thinking he was coming to India and named the First Nations "Indians." And, with Winona, I have often celebrated her. I have worked with her. Because we had fought against the biopiracy of my region's basmati rice and she was struggling against the biopiracy of wild rice. She called me to ask for ways in which you fight biopiracy. And then I brought her, I introduced her, to Slow Food. I used to be the vice president of Slow Food. And I remember joining hands because we had a basmati rice stall and she had a wild rice stall and we joined hands and said, let the Indians of the world unite. Now, let the Indians of the world unite on the Seventh Generation commitment of the kind of ancestors we need to be.

When a culture lives on the Seventh Generation, then not only does the generation thinking of the Seventh Generation to come live in a way with a consciousness leaving something good for the Seventh Generation to come. When the Seventh Generation is there it remembers all the ancestors who created the conditions of their lives. And that's why we are born in debt. We are born in debt to nature. We're born in debt to other species. We are born in debt to our ancestors. But only in that context, a debt of good action is very different. Karma.

There are two different kinds of debt. One is the debt that created ancestors, because ancestors talk of the generations that are going to come. And the other kind of debt is financial debt that takes your life away.

Monsanto's debt has been the source of indebtedness of farmers and two-hundred-fifty thousand Indian farmer suicides. So not only do you take the future away and deny the future generations from being an ancestor, but this model is taking life away from present generations. And you know I'm a physicist. We know, because of Einstein, that space and time are interconnected. We know because of Indian cosmology that *dis*, which is space, and *kalah*, which is time, are one continuum. Now when they're one continuum, what you do to your generation, through injustice and indifference, is what you do to future generations because space becomes time.

HAUSDOERFFER: *So in thinking about space and time in that way, if we break down the question of what kind of ancestor do you want to be, it's really not about how you will someday be remembered, it's really about seeing yourself as already an ancestor in this moment?*

SHIVA: Exactly. Exactly.

HAUSDOERFFER: *When you're talking about the debt to ancestors it strikes me that the phrase "ancestral instructions" that we often hear from Indigenous communities tells us that being a good ancestor means following the instructions of good ancestors. But not all ancestors are good. You've taught us, Dr. Shiva, about the creativity of women who have produced eighty thousand plants around the world and really taught us how to live in a resilient way. In Navdanya it was really empowering to be there with you on the twenty-fifth anniversary, and to see so many farmers finding hope in reconnecting with ancestral practices. How do you balance that with rejecting other ancestral practices? You know I was just in Tomic, India, and menstruating women were displaced from society, and they were not allowed in sacred groves. During menstruation they have to be with the livestock. There are empowering instructions and there are exploiting instructions. How does one choose between which ancestral practices to continue and which ancestral practices to resist? What's your ethical code for responding to your ancestors?*

SHIVA: My code is quite simple: the interconnectedness of a practice. Take for example the cow in Indian society. My mother held cows as sacred and had a cow shed and when I chucked up my job in Bangalore, you know, she said, don't even worry, don't think twice, take up the cow shed, that will be your institution. And, so, the Research Foundation for Science and Technology, with such a glamorous name, started in my mother's cow shed. When we understand our interdependence with the cow and that sacredness of the cow, it's a protection of the cow. When you separate yourself from the cow or sacred relationships like that of the cow, you separate yourself from other human beings on the basis of religion, bigotry, and hierarchies. Gandhi had said that the cow is so generous in that she supports us all her life; even when she's dead she feeds us with leather. Gandhi, the prophet of nonviolence, used to make leather, wear leather, and see it as part of the renewability of life's systems. So, when I believe a ritual is part of maintaining that web, I would follow the ancestor. Seeing interconnectedness is the test for everything. It's the test for the kind of ancestor you want to be and it's the test for artificial rituals that were built. Because we mustn't forget that you had different eras of superimposition of hierarchies. Today's patriarchy is a capitalist patriarchy that has declared a war against nature and against women. Older hierarchies were around religion. Some of them created rituals that were unnecessary. I think it's extremely important to recognize what is a ritual that holds up the web of life and what's a ritual that is as an institution a symbol of power and you shed the parts that are part of illegitimate power. Basically, you put it through an ecological test, but for that you have to have a sense of an ecological mind.

HAUSDOERFFER: *You've talked just now about the sacredness of the cow, in the past you fought for the sacredness of the tree during the Chipko movement, you've talked about the web of life just now. Are the trees of Chipko themselves ancestors, is the web of life an ancestor?*

SHIVA: Yes, of course. All beings are ancestors. The urge of any species, and it could be a plant species, it could be an animal species, it could be a microbial species, the urge of every life is not only to be a good ancestor, but first simply to be an ancestor, to have future generations. All life is based on regeneration, and yet, every step of what we call contemporary human technological progress is destroying that renewability and thinking we are the new conquerors and masters. It's like a crude Columbus version. Except the instruments are much more harmful—much, much more harmful. So, the idea of a terminator seed, that's preventing the tree from being an ancestor.

HAUSDOERFFER: *Extending beyond that, that urge of life, which I think is really compelling, thinking about your book Water Wars, when you talk about your pilgrimage to Gangotri and Gaumukh, thinking about when I visited there and a man approached me and said, "It's supposed to look like a cow's* snout, but the glacier's receded so much it does not look like that anymore." Ohm Glacier, someone approached me and said it doesn't look like an Ohm symbol anymore. Are these glaciers ancestors?

SHIVA: Everything on the planet that is part of life and that sustains life is an ancestor.

HAUSDOERFFER: *In working with farmers losing traditional connections to the land and facing the desperation of suicide, you are dealing with spiritual resilience. When you talk about that spiritual resilience it seems like, with Winona LaDuke for example, when I visited White Earth their wild rice was in crisis, and someone on the reservation said to me, "I am not Anishinaabe without wild rice in my stomach during the wild ricing moon in September." That's a very spiritual statement, but it has an ecological basis. Without the health of the land, that spiritual life becomes more complicated, and it made me wonder, in terms of this ancestor question, does the land provide an accurate barometer for the kind of ancestors we'll be seen as in the future? Does the land right now give us a way of reading the kind of ancestors we will someday become?*

SHIVA: Definitely. Because of the fact that we are interconnected. And my next book is going to be called *Oneness versus the One Percent*. The worldview of interconnectedness that leads to action of interconnectedness versus the worldview of the one percent that assumes it's just us. We need it all. The world depends on us. And trash it and trash it and trash it and they don't realize, you trash the conditions of life and you're going to go yourself. You might be creating all the systems to fly off to Mars, but it's another illusion. So, dealing with illusions is really the resilience of our time.

Spirituality is understanding interconnectedness and holding it sacred. Everything around you. Part of what has happened with the desacralization of the world, with the industrialism, which is the fossil fuel civilization, or noncivilization as Gandhi called it, is basically a truncation of all kinds, a separation, apartheid of all kinds, and one of the things it has done is separate the spiritual from the material. Our bodies were always sacred. Food, you know, a lot of my work now is on food and health, and I go back to the old texts, which say, *Annam Brahma*, food is divine. Food is divine. Food is *Brahma*. Food is the Creator. Now, in all philosophies of interconnectedness, there is no artificial divide between the spiritual and the material, because the material is sacred. And the conditions of the material, creating a condition of other materials, is well known. So, in a very ironical way, what has been described as the material civilization—in contrast to earlier civilizations, which were spiritually material—but we've defined it as only spiritual and do not care about the material. We created progress. We got to use fossil fuels. We pumped huge amounts of carbon dioxide into the air. We learned to make plastic. We learned how to kill people with chemicals.

So, it's ironic that anthropocentrism, that sees itself for humans only, and gets rid of the rest of nature, at first in our minds, and then in the real world, affects the interest of humans. In a similar way, a material civilization that ignores the fact that everything is spiritual and thus starts to destroy the material conditions of our existence or the material integrity of our lives, the integrity of the land, and the territory that gives us identity, the integrity of our food that allows the Indigenous people in Northern USA and Southern Canada to say we are the people of wild rice, just as much as when I've been to Mexico to be on the public hearings and the court cases on keeping the GMO corn out. I'm so happy they won an election where they actually finally have a leader who believes that Mexico should be GMO free, but they always begin with a ceremony. We are the people of corn. We are the people of corn.

And given the horrible hate crimes all over the world—in your country, and in India right now—I think it's very, very important to remember that so many of the categories on which leaders who use hate as capital, the categories for their using are categories of, really, a century ago, one and a half, two centuries ago. It's the result of colonialism. Cultures of the land never had flags, they had totems, but they did not have flags. Cultures of the land had territories. They knew where their territory ended and the territory of the next clan began. But they did not have boundaries written on maps. All of these are new constructs and human beings are killing each other over flags and maps. Both were created to colonize other cultures by, at that time, the privileged Europe, which then became the privileged America, which is now becoming the unprivileged America. Just like privileged human beings destroying nature become unprivileged human beings who destroy the future.

HAUSDOERFFER: *When you talk about returning to the spiritual and seeing the sacredness of the material, it really starts to feel like this is a very universal question. But is this question of what kind of ancestor do you want to be truly universal or does it need to be asked differently in different places? Do we need to ask it differently in the developed versus the majority world?*

Do I have to ask it differently as a man historically told that my ancestry is rooted in patriarchy compared to how we ask this question of women? Do we need to ask this question differently in Black communities where people can trace their descendants back only a few generations? How universal is this question? Is it problematic to treat it as universal?

SHIVA: I think fundamental questions are always universal. Particular cultures, particular contexts elaborate what that question becomes for them, in concreteness. And their responses will be particular. So, a good way to show that these are universal principles is taking care of the Earth is a universal principle because the Earth is planetary, and we are all part of her. But the Earth is also the particular place, the particular land, the particular territory where we are. So, having a sense of place does not exclude a sense of responsibility to the larger Earth at the universal level because the two are interconnected for one and the particular is always the lived reality of the universal. The problem with the universal is it's been made to be in the rise of colonialism, the rise of industrialism, the rise of this model of development, of false economies. Capitalist patriarchy was false in Europe where they were killing the women as witches and it has been false for the rest of the world, where women have been the main creators and producers yet we were told we don't work because that's the way they define knowledge, and science, and later economy. So, universal questions don't go away because they are fundamental questions about life, but they get understood and responded to within a particular context in which a person realizes that lived reality.

HAUSDOERFFER: *And I think that point about false universals is really where I'm coming from: given the history of colonialism and privilege and imperialism, who gets to decide what is universal is highly problematic. An elder in the Kakamega forest of Kenya once said to me that you cannot conserve on an empty stomach. In other words, some of these ecological questions, from his point of view, come from privilege. Is this book's question, what kind of ancestor do you want to be, one that comes from privilege?*

SHIVA: Not at all. Not at all. It's another colonialism that has allowed that person in Kenya to think conservation is at the cost of food. I've been grateful that for thirty-one years I've been given the possibility of serving the Earth and her species and farmers to evolve Navdanya and the all that has grown with it, and basically we've learned and shown the world that conservation is the basis of feeding. You conserve the seeds and biodiversity and that's where you get food.

HAUSDOERFFER: *When I visited your seed bank in Dehradun, India, I immediately noticed that it has a very ancestral feel to it. In your choice of murals, just the style of construction, you evoke the feeling that it's a sacred space. I walked in and each seed revealed thousands and thousands of years of, really, scientific method and trial and error, and what works in that place and that climate, and ancestral instructions are in the seed, which is also the ancestor. Even further, the fight for that seed is how you become a good ancestor and ensure that there are future ancestors. I think that's a really hopeful and compelling vision. In closing with such hope, what would you want the reader to pause and contemplate before answering this ancestor question?*

SHIVA: When they pause and contemplate, to be a good ancestor, you have to know you're significant. Your life has meaning and you have something to give. We have been so badly trashed that most people feel, "I'm nobody." Because that's what the system has told us: You're nobody. You're disposable. No, each, the tiniest microbe has meaning and significance. You have meaning and significance. And go into yourself and think of what is meaningful to you. Because that, then, is what will give you a sense of what you want to hand over to the future and answer "what kind of ancestor do you want to be?"

POEM

Your Inheritance

Frances H. Kakugawa

This Earth you call home
Was not created by chance.
There was no magic wand
That preserved man's inhumanity to man.
There was no magic wand
That kept all life in oceans, air and land
Free and clean
So you could breathe, swim, drink, live
Without fear or peril.
There was no magic wand. No.
It was I, millions of I's
Generation after generation
Who preserved, restored, and renewed
This legacy now in your name,
Stewards of the new world.

✶ 6 ✶

Seventh Fire

Neesh-wa-swi'ish-ko-day-kawn arose and said:
 In the time of the Seventh Fire an *Osh-ki-bi-ma-di-zeeg'* (new people) will emerge,
  they will retrace their steps to find what was left by the trail . . .
 The task of the new people will not be easy. If the new people remain strong . . . there will be a . . . rekindling of the Sacred Fire . . .
 It is at this time that the Light-skinned Race will be given a choice between two roads. If they will choose the right road, then the Seventh Fire will light the Eighth and final Fire—an eternal Fire of peace . . . If the Light-skinned Race makes the wrong choice roads . . .

EDWARD BENTON-BENAI, *The Mishomis Book*

POEM

Time Traveler
Lyla June Johnston

Time traveler running faster.
Warrior is born.
Battle to be won.

Past trauma, future hurt.
I'm a child of the dirt and
I'm ready to give birth.

Planting a dream.
Panting, I breathe.
Running towards the future
with a handful of seeds.

Stronger than greed.
I am stronger than hate.
I stand under the shade
of trees planted so long ago.

A product of ancestral love,
I'm here because my elders
danced in the sun.

They would give it all up for us
and from day one it was
practiced like religion

to prepare for the ones
to come.

We are here
to give all our love
to the ones unborn.

But this is insane.
Living for fame.
Living for the next quarter,
profits and gains.

You forgot love.
You forgot truth.
You forgot how to live for a time
beyond you.

It's not about you.
It's not about you.
It's about the song that is
traveling through.

It travels through time.
Singers will die
but the song lives on
through matrilineal lines.

We are here
to give all our love
to the ones unborn.

Open your eyes.
Open your heart.
Draining aquifers before
they can recharge.

We're not in charge.
Nature's in charge.
Look to the stars
remember who you are.

Stay humble or fall.
We don't know it all.

And we are not exempt from
natural law.

Live selfishly
and the structure will fall.
But if we live for those unborn
then the song will go on.

So before you take a book off the shelf,
take a look inside yourself.
Answers come to you at light speed.
I'm searching for knowledge
I can't find on a newsfeed.
Knowledge found through intuition.
Knowledge found through fasting and dancing.
This ain't superstition.
It's ancestral tradition.

Throw me the spear of wisdom—
sharpened and sunlit.
I'm giving my life to the oneness.
I'm a warrior. I'm sun-kissed.
I'm armed but I'm harmless.
Protecting cycles of rain and cycles of snow.
Fighting for children whose names I will never know.

I look up and read the messages written all across the sky.
Messages telling us that it's time to evolve or die.
It's time to live this life right.
So that when our children look back,
they look back with pride.

ESSAY

Seeds

Native Youth Guardians of the Waters 2017 Participants and Nicola Wagenberg

Our seeds are our children and at the same time our ancestors.

GUARDIANS OF THE WATERS

SECTION I

In the summer of 2017, a group of 20 Indigenous youth, aged 16 to 30 years, participated in the Cultural Conservancy's Guardians of the Waters Youth program. The youth represented diverse tribes and nations from Native California, North America, and around the globe.

Throughout our program, we use expressive arts as a way for participants to reflect, process, and integrate what they learn. In one of the activities toward the end of the program, we asked the youth to answer the question: *What kind of ancestor do you want to be?* They could choose any art form to answer the question. Following are some of their creative responses.

*

Corn is the ancestor whom I most want to learn from, whom I most want to be like. Each of her seeds holds the memories of every generation past—every planting ceremony, every mutation, every rainy season, every drought, every plant generation before her, every human relation who has nourished her—and all of these memories are the origin of her deep resilience.

I am relearning my ceremonies, reconnecting to my communities, planting our seeds, and fostering relationships that have not yet been lost, because like corn, I want to be the type of ancestor who carries all of the other ancestors with me.

[Maya Harjo (Choctaw, Quapaw, Shawnee, Muscogee Creek, Seminole, Delaware) is a 28-year-old queer Native farmer and educator who grew up in Los Angeles, regularly visits her tribal communities in Oklahoma, and currently lives in the Bay Area, but who finds herself in seeds.]

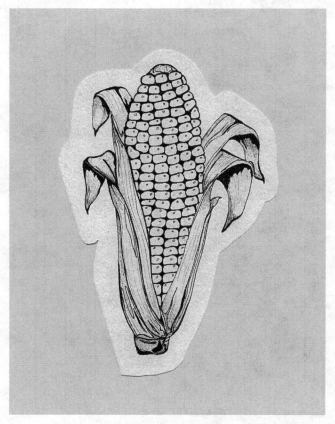

FIGURE 1 Maya Harjo

*

What Kind of Ancestor Will I Be?

Love, hope, passion, pride, art
Sex, breath, perseverance, empathy
Gratefulness, honor, reckoning, redemption,
Rendezvous, remembrance, humility, sincerity
Androgynous & ambiguous
Remember all of my faces, but call me by my one name
walking contradictions. Legacy of oxymorons.
I never want to be explained, only understood, only felt.
Language learner, reclamation to lineages lost.
The Reconquista lives in me, fueled by a tortured tongue.

I am the seeker. The inventor—the forgiver.
I am the pardoning of my gunman.
The kiss shared between us before I became his victim
I want to be the author, editor, and publisher of my own story.
I want to be proud in every life, in every realm of being,
regardless of how much it hurts.
I will be the force that pushes back when gravity demands we surrender.
The force or name you call the action of bouncing back.
When you remember me, think of love.
Remember my laughter like the spirit of song.
One of many Buddhas.
Let them honor my gender in completion and all of its complexity.
I am the first breath of air that fills your lungs
as you gasp and spring up from under water.
I am the creator who spent half of their lifetime searching
for something that was begging to be born
from my own hands, my own heart, and my own smile.

BENNY AVALOS

[Benny Avalos is Queer, first-generation Chican@, revolutionary.]

*

What Kind of Ancestor Do You Want to Be?
A Plantcestor!

I want to be the ancestor
who planted the three sisters
we wishful thinkers . . . Why? Because we are resisters!
I want to carry the medicine, to just run with it and be level headed with it

Take care of you, take care of me, all that you gotta do, plant them seeds

The acorn on the oak tree
I want to be . . . I want to be . . . resilient like a cactus staying sharp without acting
I am loving and free: squash, maize and beans

I got the heat
the water that I drink splashing on my feet

the air that I breathe
the wind on my leaves
I got the key, you can come with me or you can leave

SONG BY ANTONIO FERRER

[Antonio is Mexicano, white, and Cherokee, an avid gardener and lover of plants.]

*

I drew the moon, mountains, cactus, little seedlings. The questions written around the border of the drawing read, "What kind of ancestor do you want to be? What are you doing for the future seven generations? ¿Que tipo de ancestro quieres ser tu? ¿Que estas haciendo para las siguiente siete generaciones?" Putting this phrase in both Spanish and English is an ode and acknowledgment to the double-lens that I have on the world because of the languages that I speak—without which I would not have the understanding of the world, or myself, that I do today.

On the mountains it reads "Mamá Montaña dame fortaleza. Ancestros, sean me guia" (Mother Mountain give me strength. Ancestors, be my guide). Not only do mountains symbolize wisdom and great strength, but my ancestors, the Purépecha peoples of Michoacàn, settled in and around the mountains of Michoacàn. Turning to the mountains for strength and guidance is like looking into the past and asking my ancestors for strength and guidance.

In the center of my drawing is the full moon. Written on top of the moon is a prayer that reads "Nana Cutzi, dame vida, dame luz . . . Cuida mi matriz y aguas" (Moon, give me life, give me light . . . Take care of my womb and waters). In Purépecha mythology Nana Cutzi is the Goddess of the Moon. Being that the first element that every human being starts off in are the waters of their mother's womb, it is important for me to pray to my ancestors to protect that which is the essence of me, my ancestors, and the future seven generations that are to be born of my womb water. In addition, the moon is symbolic to the waters in my womb because it is the full moon that makes the oceans' tide rise and fall, just as it is the moon who tugs at my waters and makes my womb dance. Look up to the stars of my drawing and amongst the stars you will find the phrase that reads "haceme brillar" (make me shine). Just because the night is dark does not mean that we cannot find our light to make us shine. Whether that be the sun by day or the moon and stars by night, when connected to my ancestors, I will always find the light to shine.

Turn away from the sky and now to PachaMama, the Earth. Here you will find nopales. Nopales, soft and sweet enough to enjoy this delicious and nutritious ancestor food, yet tough enough to defend itself with its large thorns from those who do not know how to handle them with care. "Nopales enséñame a defenderme de lo que no me alimenta" (nopales, help me defend myself from that that does not allow me grow). I ask the nopales to help me shed away all that does not nurture me, all those who do not know how to handle me, and all of that energy that will not serve me.

Everything in my drawing has life. The nopales have the little espinas—if you touch them, they'll poke you. Yet, they bear the delicious tuna that when thirsty, will quench your thirst. Tonantzin, ground me, PachaMama, ground me. The little seedlings remind me that patience and trust in myself is essential to my growth, and thus to nurturing those generations that are to come of me. These are different prayers that I send to my ancestors so that as an ancestor myself, I can embody and reciprocate those prayers unto my children,

FIGURE 2 Jessica Garcia Ibarra, pen on paper

FIGURE 3 Nance Puc, oil pastels on paper

grandchildren, great-grandchildren, and those generations of children that I will one day meet when I am a mountain ancestor myself. For now, I must never forget that my ancestors are all around me as the elements on Earth that I sing to and say little prayers for.

[Jessica Garcia is a 25-year-old, queer, muxer, Indigenous to the lands and waters of Michoacàn, Mexico. Raised in Southern California, Jessica now resides in the Bay Area.]

*

We connect with our ancestors through nature (the land), cultivating it. I have not thought of them [ancestors] a lot like I have through this program. Loved the "what kind of ancestor you would want to be" question. I never saw myself as an ancestor passing through life.

[Nance Puc (Mayan) is 28 years old.]

*

LUNA'S FREE STYLE WRITING

Let's say I make ten billion dollars throughout my life and I have two kids and each one has five billion. You know, then they have two kids of like 2.5, 2.5 and then eventually every single person on this planet will have a penny from my life. And I think how much does a penny equal in this world. Yet how powerful it is to know that we're all related and we wouldn't want our daughters fighting our daughters' daughters. Yet we fight.

It helped me realize acquiring material wealth as a central factor to my genetic preservation is not the best way to go about things. It's all about investing in each other. Like my brother, he has questions with his sexuality and stuff like that and just me painting my nails, wearing lipstick and stuff, it's changing the world you know. You can't buy a more accepting world. I am making the world more accepting. And you know it helps my brother. It helps my future generations. Like the movie Interstellar, what is time and love and all this?

Ideations of a cosmic society and the nonlinearity of time. You all will see Luna (the young eyes of the divine feminine) again as you will see Yama (the cosmic artist). We will be brothers, lovers, mother and son, business partners and enemies. You are my ancestor and I am yours. I'm an ancestor now, my life is encoded in my DNA. Whether or not I replicate through the physical or just epigenetic enrichment through psychosocial realms. So this question of what kind of ancestor do I want to be? Not having the elders I was looking for getting here led me to become them. I can see my existence having an impact right now. And I love myself and I continue to fall deeper every day. I just want to say you are all beautiful people and as much as an ancestor I can be, I already am. You guys are beautiful people, aspirations of who you want to be—you guys are already, you know. I don't know. Beautiful.

[Luna Lucich (Choctaw) is 23 years old, Two-spirit.]

*

Indigenous youth of today are awake, they want to reconnect with traditional knowledge and practices, and protect what they have for future generations. They are the water protectors, seed savers, language learners, decolonizers. They are agents of creation, change, and liberation. They are challenging colonial boundaries and redefining gender and sexuality. They are expressing new Indigenous identities and forging contemporary intertribal collaborations.

Our lives and future generations depend on them. Let's support them to be Guardians of the Waters, Guardians of the Earth. Youth of today, ancestors of tomorrow. We stand with you.

10 key lessons from the Native Youth Guardians of the Waters 2017 Program

1. We can connect to our ancestors through our food and seeds.
2. Water is the sustainer of life and cultural traditions.
3. Preservation of seeds is the preservation of a Nation.
4. Loving, nurturing, and affirming communities exist where one can fully be themselves.
5. Saving our soil, our water, our animals, our plants, and our seeds is essential to overcoming the struggles of our people.
6. Listen to plantcestral love.
7. Traditions and cultures can and do evolve.
8. Reconnection to one's culture is possible at any time in life and space.
9. It is important to listen with all of our senses and feelings. Or else we can miss messages from nonphysical / nonhuman relatives.
10. Tending personal waters and plants is a metaphor for how we treat our immediate and communal relations. This requires compassion, intention, and nourishment.

SECTION II

What are we doing for the youth of today, especially Native and Indigenous youth? What do they need in this time of climate chaos, uncertainty, inequality and yet so much potential? How do we support them to reconnect and activate in their own ways the knowledge and practices of their ancestors so they can understand their innate connection to land and become guardians of the waters?

With this in mind, The Native Youth Guardians of the Waters Program (GOTW) was born in 2013 out of a long-term vision of the Cultural Conservancy. GOTW is an innovative program that provides an immersive Indigenous-based educational experience to Native and Indigenous youth in the San Francisco Bay Area. It is one of the programs of The Cultural Conservancy, an Indigenous-led nonprofit organization based in San Francisco with over

30 years of work with Indigenous communities. The program facilitates cultural, spiritual, and emotional healing through the revitalization of personal and interpersonal relationships to Native lifeways, arts, and traditions. Youth connect with Native artists and knowledge holders, learn about Native foodways and Native water traditions, participate in talking circles and group process, engage with traditional and new art forms, and immerse themselves in nature. Participants explore the connections between Native identity, ecological and cultural health and healing, and community engagement.

Eco-cultural health is the understanding of and respect for the values, beliefs, and practices of the people in relationship to the land they come from. This means having an accurate and productive sense of one's cultural identity, which in Native communities includes knowledge of one's language, creation stories, foods, arts, and ceremonies. It connects us to the ancestral legacy and traditional knowledge of our ancestors so that ultimately it contributes to a greater sense of self, capacity for agency, creativity, and transformation. Ultimately, this enables us to have a deeper relationship to the land and helps us be better guardians of lands and waters.

In Native cultures there is a long tradition of honoring humans' role as guardians, with a fundamental responsibility to an ethic of guardianship toward all living beings, including land, waters, and stars.[1] In this program, youth are invited and guided to understand what it means to become guardians of the waters, the waters of their inner ecologies and the lands they inhabit and come from. As one of our youth from this past summer describes:

> Being a guardian means I have a responsibility to Mother Earth, Father Sky, the wind, the fire, the water spirit, to love it by protecting all the sacredness. It is my duty to honor my ancestors. I can no longer stay idle, silent, and indecisive. I have all this knowledge gifted to me. Restore our planet to a healthy state. I will transform the disease/illness of this planet and bring balance.

In our program, youth reconnect to and learn within the cycle of planting, growing, harvesting, cooking, and eating organic healthy and culturally appropriate Native foods. For many of them it transforms their eating habits and improves their health. Youth also develop emotional awareness and capacity. Through talking circles and expressive arts the youth get to share who they are, where they come from. Many feel and express the pain of historical trauma and the destruction of the lands and disconnection from ancestors.

One of the most important parts of this work is that participants learn by doing. They embody the knowledge that is presented to them by learning basket weaving, traditional songs and dances, and how to carve salvaged native trees, among other hands-on activities like planting seedlings. They explore how their ancestors farmed, "tended the wild," and cared for their homelands for millennia grounded in a knowing of the critical interdependence of people and the earth.

The youth learn from many Native artists and elders, Guardians of their culture, who teach about plants and animals, finding home in the stars, and caring for watersheds. Youth learn how the native plant tule acts as a water filter helping conserve and purify our waters; they learn how to harvest its shoots and roots for food and how to weave with it after it is harvested and dried. They participate in a water ceremony with Ohlone women leaders by the historical waterfall in Indian Canyon, the only federally recognized Ohlone territory. They sing with a Miwok elder at Kule Loklo at the Pt. Reyes National Seashore. They learn about the devastating effects of dams on our rivers and the salmon restoration efforts in Mt. Shasta from a Winnemem Wintu young leader.

Every spring for the past five years, we have put out the call to youth across the Bay Area who identify as Native American or indigenous to other parts of the world, to join our program. The program takes place in various outdoor spaces: urban gardens, urban native plant and water restoration sites, rural farms, the ocean, lakes and rivers, mostly in Ohlone, Coast Miwok, and Pomo territories. For many it is their first time leaving the city, camping under the stars, and swimming in a lake. They shed layers of protection and are able to be more fully who they are. Many participants express how our program helps them connect with their true self. It gives them a sense of purpose and impacts what they choose to do next: organic gardening, Native foodways cooking, water restoration, and outdoor education. They learn that the outdoors and nature are not separate from themselves but that they are a part of and belong in nature.

> I loved the ability to build community, to share and hold space for each other. The amount of love that was felt and developed brought me so much joy. Also, the ability to have access to such incredible teachers, food, and knowledge. I loved the connection with nature and the time we spent outside. I also really loved the rituals, building altars, sharing medicine.

To learn more about our program visit: http://www.nativeland.org/youth

ESSAY

Onëö (Word for Corn in Seneca)

Kaylena Bray

I look at corn and I see an ancestor. For as long as I can remember I've been surrounded by the taste and smell of Seneca white corn grown by my dad and grandpa. This corn is called Onëö'gan in the Seneca language. Our farmland is located in the thunderous Western New York region, and made up of rich clay soil that seems to plaster every footprint and tractor tire like individualized molded art. The soil's moisture-rich crevices attract an abundance of millipedes and worms deep enough to quench a young girl's underworld exploration, and every now and then a scraggly arrowhead reaches the surface after a hard rain. On a lucky day, there will be an eagle feather, majestically strewn atop the soil by a passerby heading toward the creek alongside the cornfields.

These memories punctuate my childhood, and to this day, each time I take a bite of corn soup these stories are reawakened viscerally and subconsciously. They animate my understanding of corn both as an ancestor, and as an emblem of knowledge passed down by countless hands of farmers, seed savers, youth, and elders. In coming to understand the world of corn, I've seen how its spirit vibrates within the people it touches, and I recognize this spiritual element as a binding force that gives face to the legacy of movement adaptation, and respect that corn has endured over its history as an ancestor. Corn spans a longitude and latitude of movement greater than any other domesticated food source. It is remarkable to think about the extent to which farmers and seeds have endured and adapted to the driest of desert landscapes in the Southwest and across Turtle Island to the thunderous and snowy regions of the northeast.

In truth, it is this relationship between farmers and their corn that characterizes the bond most central and poignant in my understanding of what it means to maintain ancestral relationship. Farmers must care for their corn as it grows through the many stages of development, from green infancy to the formative milky stages of teenage growth, and onto the strengthening of mature stalks. Much like caring for a child, a farmer must protect the corn from potential hazards, and ensure the regeneration of seed into the following generations of growth. Alongside the ancestral lineage that binds people to corn, the affixed relationship between farmers and corn hinges on the day-to-day bond that forms in caring for a plant that relies on humans for life. And in turn, has enabled human life.

How beautiful to watch the unfolding of a relationship that binds plants and humans. The more time I spend in dense urban landscapes, I notice myself more consistently forgetful of the intimate bond between human and plants. It is why these memories of corn have become increasingly important for my grounding in ancestral connection, and it is by examining my relationship to corn that I find it possible to explore the intricate threads linking my physical, spiritual, and cultural existence to a greater understanding of human and nonhuman relationship.

In May 2016, I began a journey, starting at Onondaga Nation and through New Mexico, before returning to the New York region of Onondaga Nation. My dear friend and talented videographer Mateo Hinojosa and I went about visually and viscerally documenting the dynamic interplay characterizing the intimate farmer-corn relationship. We sat with corn growers of different ages and Native Nations, and asked questions about the nature of corn in cosmology, corn as part of everyday life, and its significance in understanding human responsibility in the world. On our first trip, we interviewed farmers from the Six Nations of the Haudenosaunee Confederacy, whose legacy with corn and seeds extended multi-generationally. Our conversations shifted between the practical and the humorous, inevitably landing on the emotional threads linking the history of ancestral sacrifice and endurance to a deep recognition of its existence within and alongside seeds today.

On one visit to the home of Faithkeeper Oren Lyons, we spoke about the concept of seed sovereignty. "To be able to feed your people, that is sovereignty," he spoke. As a Haudenosaunee woman, I recognize that the seeds of my ancestors—the Seneca white corn seeds I've grown up eating—carry a strong legacy connected to sovereignty. This sovereignty relies on the health of our ancestral knowledge systems, the health of the land, and ultimately, the health of our people. As my conversation with Oren steered toward the state of environmental change, we came to a growing realization of the need to acclimatize food sources in the face of rapid change. How then, I wondered, would seeds need to be understood, exchanged, and grown in the context of these pending environmental extremes, and what were the responsibilities given by our ancestors through the seeds?

While in New Mexico, I understood just how deep the roots of human relationship extended to the environment, especially to the rains. The expanse of dry red desert landscape made the absence of water especially palpable in the dusty cornfields where the stunted corn stalks receive very little water during the growing season. On a warm afternoon in August, I met with farmer Jim Enote for an afternoon of farming and discussion on his farm in Zuni. We took to irrigating the rows of corn and vegetables by siphoning water from a shallow ditch shared by the farmers in the area. As we carefully flooded the plot, I realized what Jim meant when he said caring for these seeds was like caring for children. There we were creating a comfortable home for the corn, making sure the soil was nutrient dense and fluffy, properly aerated with care so the plants could grow firm roots. Not to mention, ensuring their life proliferated with water brought by old technologies of irrigation passed down by generations of farmers. In undertaking the careful process of irrigation, I realized the extent to which these intricate knowledge systems governing the care of these seeds are equally as vital to the act of resilience as the seeds themselves.

My final trip in this corn journey took place again at the Onondaga Nation. This time, I was there to take part in the documentation of an extensive seed collection gifted to the Onondaga Nation by seed collector Carl Barnes, a renowned Cherokee farmer who spent his life growing, collecting, and continuing a legacy of corn farming on his land in Oklahoma. The seed collection is currently stored in a warehouse on a plot of farmland overlooking a sprawling hill of old apple orchards. Without fail, every time I walk into the seed house I have the distinct feeling that I am surrounded by a group of elders saddled with the knowledge of lifetimes past. It's become a strong reminder for me of the relationship that bonds humans and plants as student and teacher, as elder and youth, and as ancestor and grandchild.

One of my favorite seeds from the collection is the grandfather pod corn. Each seed on the grandfather pod corn is individually wrapped in tiny husks that give the appearance of a leafy mass of fossilized seeds from a dinosaur era. Looking closely, I can catch glimpses of marbled corn peeking through the thin sheaths of the seeds like hidden gems waiting to be uncovered. In truth, pod corn is one of the more mysterious corn varieties. Scientific debates abound as to its true genetic origins, and little documentation has been uncovered around its spiritual, medicinal, or food value as a plant. Upon closer look at its sheathed exterior, it is apparent that something unique has allowed this corn to be here, given how difficult it would have been to uncloak each individual kernel to gather enough food for an entire village or community. The mysteries surrounding pod corn are just as cloaks of its kernels, and the depths of this unknown story seem tightly interwoven with the profound symbolism of its survival, and human influence on corn.[1] Through this lens, I try to comprehend how far back the relationship between humans and corn extends, beyond time and place, and as part of countless stories of cosmology and creation.

The ancestral lineage carried through seeds has become a tangible reminder for me of what it means to be an ancestor. The grandfather pod corn was given its name for a reason. It acts as an ancestral grandfather. It is a source of strength and resilience that carries its influence in unknown and lasting ways, and watches protectively. I feel a similar source of influence from the corn I grew up eating, *onëö'gan*, and it strikes me how unknowingly yet persistently I've had this connection my entire life. To this day, when I eat white corn in soups or boiled bread, I think of what life must have been like for my ancestors, and the strength and resilience needed for this corn to be here. I think of the French Expedition of 1687 where they burned half a million bushels of white corn in a raid designed to wipe out the Haudenosaunee people at Gannagaro, present-day New York. Despite these attempts we are still here, and the corn is still here. There is a sign displayed prominently in the seed house of this extensive and ancient corn collection. It's a framed photo of corncobs and imprinted seeds that reads,

"They tried to bury us. They didn't know we were seeds."

ESSAY

Landing

Oscar Guttierez

> I hope you end up liking Huntington Park.
> Why wouldn't I?
> Cuz sometimes the air smells like pan dulce and others say it smells like chemicals.
> Or like Las Cuatas say, "It smells like ass."
>
> MOSQUITA Y MARI[1]

My involvement in the environmental justice movement began in 2008. I would walk from school over to the offices of Communities for a Better Environment (CBE), an environmental health and justice community organization. The office was located in between Vernon and Huntington Park in southeast Los Angeles County on the traditional homelands of the Tongva people. I grew up in Huntington Park, but one could cross over to Vernon, which was exclusively industrial with about 112 residents who resided on one small block. I did not live in the city of Vernon, but the city definitely lived in me. It made me ill every morning and almost everyone in Huntington Park understands this. No, the problem at the time was not the dangerous levels of lead and arsenic that were entering our bodies, although that was and continues to be an issue. Instead, it was the stench. Aurora Guerrero captures it best in her film *Mosquita y Mari* when the two main characters overlook the Los Angeles River in dialogue. CBE would host an event called the Toxic Tour, which consisted of a bus that drove around the city exposing the largest polluters in our community. Toward the end of the tour, one is exposed to the source of the stench. A wall and a small gate stand in front of the meat rendering plant where a large pile of animal carcasses is out in the open. At this point in I have had multiple interactions with the rendering plant and every single time I am reminded about the harsh realities this type of work brings forth. There is a striking memory contained in that smell that reminds me that the work is far from over. The joke of Las Cuatas is reminiscent of conversations I often had with others in the community. But the harsher realities lie in being subjected to a pile of animal carcasses. What does it mean to be constantly close to death?

In a recent visit back to Huntington Park, I visited the offices of CBE, where I encountered a photo that had been hanging on the wall since before I even started working with the organization. The photo shows community residents who took on one of the first campaigns with the organization in the late '90s that consisted of the removal of a large mountain of concrete rubble that was left over from the Northridge Earthquake. I was told the majority of the members in that photo had joined the spirit realm. While it is undetermined whether their death was caused by the health effects that were results of La Montaña,[2] I am forced to think about what those people wanted for me and how the legacy of their work prompts in me the question, what kind of ancestor do I want to be? There is a particular honor that is contained in this question that activates the memory of my ancestors. I want to think about this question alongside something my mother always told me. She would often say something along the lines of "you are earth and earth you will become, again." Hence, while I answer the question of my responsibility to the future generations, I also choose to leave something for my own ancestors too. Those who have and will become earth, again. I hope to further examine how death had taught me what I know about ancestry and how my hopes for the next generation are in an assemblage of the past, present, and future legacies of our stories.

ANCESTRAL (RESPONSE)ABILITY

My ancestors have left me the tools to move forward, but also take necessary steps back. While I have a responsibility to my own community, my larger responsibility engages a deeper understanding of what it means to work in relation to the land. Hence, I ground this relationship in the work of Robin Wall Kimmerer, who asserts, "Our relationship with the land cannot heal if we do not hear its stories. But who will tell them?"[3] In beginning to answer the question of what kind of ancestor I want to be, it is important that I locate what kind of ancestor is needed in the world today. While we see the degradation of Indigenous land through the current administration in their assertion of everything from pipelines,[4] to the threats to sacred sites like Bears Ears[5] and the lands of the Winnemem Wintu tribe through expansion projects of already disastrous dams,[6] it is important to locate how we understand our responsibility to Indigenous land everywhere. I truly believe we are at a turning point in the world. While these threats on Mother Earth seem constant, so is the resistance. Our ancestors are powerful and the work of protecting Mother Earth is at a crucial point. There is something incredibly moving about doing this work knowing you are here because of the work of your ancestors. As Sandy Grande asserts, "In a time when the dominant patterns of belief and practice are being widely recognized as integrally related to the cultural and ecological crises, the need for understanding other cultural patterns as legitimate and competing sources of knowledge is critical."[7] The knowledge of the original caretakers of the land is central in a future ancestry, and rightfully, without our ancestors, there is no future.

The current movements to protect Mother Earth reveal commitments to our ancestors. While, for many of us, tracing our ancestors has become a task that is sometimes impossible because of the interruptions of colonialism, I continue to honor them through the defense of Mother Earth. After all, the land is our largest ancestor. As Mishuana Goeman states, "Land in this moment is living and layered memory."[8] The assertion that land is living and layered memory activates an understanding of the land as our largest informant and our most sacred connection. In this framework is where I place a departure as my responsibility as an ancestor. While Kimmerer asserts the need for someone to tell the stories of the land, the larger aspect of this question is making sure that we are listening. Hence, the future of my ancestry depends on the "original instructions."[9] Although I am largely concerned with the future, I must affirm that the concept of a future ancestry is primarily tied to an understanding that ancestry is not on colonial time frames. While we can understand ancestry in the future, we can also understand it in past, present, and outside of time completely. So, to understand my own future as an ancestor, I must understand the shape of my impact[10] in this moment and how future generations can carry on the work that honors the earth. Additionally, as I stated previously, ancestry cannot simply be understood in the human form. Ancestors cannot only be contained in such forms.

I understand my work with the land largely as a conversation with it. The current conversation with the land is amplified, and our responses are crucial. Part of my conversation is an agreement that, as my mother asserts, I will become earth. While I largely reflect on the responsibilities set forth as a future ancestor I also assert that sometimes, my instructions and duties as an ancestor are laid out before me.

I WILL CRUMBLE INTO THE EARTH

My mother had her deepest connections with the earth when she mourned. My mother saw my grandfather transition into the spirit realm and when she came back from her motherland she was in a silence none of us had ever seen her in before. She went straight into the small garden she had made for herself in the front of our apartment and began doing yard work. Sometimes my mom would dig so deep I thought she was attempting to find the connection to her father in the soil. I remember her being deep in the soil when I would get back from school and sometimes losing her in the middle of it all. As Deborah A. Miranda would describe in her own memory of her father, "Sometimes it was hard for me to tell where my father ended and the earth began."[11] Maybe this was what my mother was talking about when she said I would become earth. My mother was never the same from that day on and I don't know if she ever found my grandfather in the middle of that plot of soil. Maybe my own desire to put my hands in the soil is in finding my grandfather to finally ask him the questions I never got to ask him. The following summer the most vibrant chiles[12] I had ever seen and tasted sprouted from that plot of soil.

To do the work of tending to our Earth Mother is to do the work of tending to our ancestors. They exist in our foods and allow us to have our abundant harvests. I often think of the resemblance my grandfather's skin had to that plot of soil during that warm summer day. My last memory of my grandfather was on that plot of soil my mother had deeply dug into. He was sitting there with my mother and I greeted him and walked away. It was as though he never left, just crumbled into the soil. When my mother said, "you are earth and earth you will become, again," she was giving me directions on how we will find each other again. My mother teaches me how to be an ancestor. My ancestral teachings have a strong central connection to the land. I never thought I would learn so much about gardening when I grew up in a tiny apartment in Los Angeles. There was nearly no space to grow food, but my mother found a way. My mother always said that there were certain foods she could no longer make because they reminded her too much of my grandfather. Much like conversations with the earth, I deeply value the type of ancestry that comes from foods. Foods have the power to make us remember them even when we don't have any memory with them. As Enrique Salmón asserts, "Food itself activates in us encoded memory that reminds us how to grow, collect, and prepare food. The land and food then become sources of knowledge and history."[13] The knowledge that food produces in the self is a place where ancestors are often found. Consider these conversations, a visit to your ancestors.

I have a wish that when my body becomes earth, the soil will sprout chiles like they did for my grandfather. When I crumble into the earth I dream that the seeds of my fruit will be spread into fields that are larger than my body. The land and my body join in a joyous celebration that honors my past, present, and future. This celebration is also yours.

FUTURE ANCESTORS

Thing is, I don't believe in ghosts. But I see them all the time.
SHERMAN ALEXIE, *You Don't Have to Say You Love Me*

I see ghosts all the time. I see them in the meat rendering plant, the Oakland crematorium,[14] and the Shellmounds[15] in the Bay Area. I see them all the time and I can't get away from them. There is something frightening, yet settling in seeing yourself reflected in them. Sometimes I see my ancestors in photos, other times in my mother's foods, and sometimes I see my ancestors in each of us. The current moment we are in is vital. We walk amongst powerful future ancestors that carry the strong legacies of generations. More importantly, we are seeing a generation of youth that are empowered and proud to practice traditional ways of being and have confidence in their worldviews. The current attacks we are witnessing on Mother Earth urge our work to move into larger depths. It forces us to think about our movements and the need for decolonization. As I think deeper about the ancestor I want to be, I think about the ancestors that hold me to this work and continue to believe that I, much like them, can move mountains. I am nothing as an ancestor without the teachings of future generations. Future generations are essential in activating the legacies of our ancestors.

The ghosts of my ancestors follow me and I am indebted to them for their sacrifices and hard work. They remind me of the work I need to do on the ground and my obligations to the next generations. While I continue to engage in a making of worlds,[16] I assert that I have a deep confidence in being an ancestor that tells the stories of the land, that holds a responsibility to the original caretakers of the land and that trusts deeply in the dedication it takes to defend and listen to the stories of the land. I am at a point where I am ready to walk with the ghosts in my life. They have the teachings for my own haunting.[17]

ESSAY

Regenerative
Melissa K. Nelson

I am a descendant of the Lynx moving along the edge of forests. I am a descendant of Buffalo hunters on the Northern Plains. I am a descendant of wild rice and sweet lakes. I am a descendant of cranberry gatherers in the Turtle Mountains. I am a descendant of earth divers and round dancers. I am a granddaughter of Mary Sauvage, Mary Cree, and Mary Caribou, my Indigenous foremothers who were given Christian first names and kept tribal markers as surnames. *Boozhoo Nokomis!*

I am a descendent of French Voyageurs seeking soft beaver skins for money. I am a descendant of women who married bears and made moccasins. I am a descendant of traders and merchants carrying canoes loaded with animal pelts across portages in the prairie. I am a descendant of fiddle players and jiggers. *Bonjour Grand-meres Poitra, Malaterre, Amyotte.*

I am a descendant of boarding school survivors and Michif-language speakers, "half-breeds" and passers. I am a descendant of winter starvation survivors and rabbit trappers. I am a descendant of relocated urban Indians raised by Catholic nuns and frozen juneberries. *Tansi Nokom!* I am a descendant of the Sixth Fire.

I am a descendant of Scandinavian women who made mead and went to battle with metal shields. I am the descendant of Norwegian farmers growing barley in North Dakota throughout the twentieth century. I am the descendant of Protestant preachers and elementary school teachers. I am the granddaughter of the Nelsons, Evansteds, and Semlings. *Hei Bestomor!*

In greeting my ancestors, I recognize that I am touched with all of the stories that they carried as human beings on the Earth: their seeds, waters, traumas, legacies, worries, pains, victories, and love. I am tied to them physically and metaphysically as my ancestral roots. As the late great poet John Trudell would say, we need to recognize our true DNA, our Descendants-N-Ancestors. In the Anishinaabe language the word for ancestor is *aanikoobijigan*, which is based on a root word meaning "string or tie it together." To be an ancestor is to be tied together throughout the generations. But what is it that ties us, genes, behaviors, blood, languages, identities, breath?

As my Tongan elder friend Emile Wolfgramm often asks me, "Are your actions worthy of the conspiracy of your ancestors?" What did they have in mind for me and the future of our kin? Ancestors give us life, so that is first and foremost. We wouldn't be here without them. They provide identity and history. Through their stories of struggle and survival, they also give strength, perseverance, and courage. My ancestors have also given me the ability to embrace a diversity of perspectives, representing both the colonizer and the colonized and the messy, mixed spectrum between. In this reflection, they help me think about what kind of legacy I can leave for future generations and what kind of ancestor I want to be. I wonder what is it that gets transmitted and why. Is this a deliberate act like planting a domestic seed in well-prepared ground, or is it a wild process, like big leaf maple seeds spinning through the sky after a big gust of wind? Trickster consciousness tells me, of course, that it is both and neither.

I want to be an ancestor who "remembered to remember" and who worked to restore the fragments of Turtle Island knowing and being out of the wreckage of colonial collisions and monocultural assimilation. This is what it means to be a descendant of the Sixth Fire. There was dislocation and ethnocide, cultural rupture from boarding school, stolen and poisoned lands, and enforced poverty and dis-ease due to oppression and racism. The prophecy of the Fifth and Sixth Fires said there would be much pain from so many relatives pushed off the Red Road and forced into the Split-Head Society.[1] Too many dying from the "soul wound" of colonialism or disappearing in a bottle of oblivion to self-medicate from the intergenerational grief passed down.[2] These are some of the legacies of colonialism that I inherent. These mark my body, mind, and psyche in subtle and overt ways: genetics and epigenetics and Adverse Childhood Experiences (ACE) and endocrine disrupters. All real and measurable on the geography of Native presence.

These traumas I inherit along with profound resilience, a faith in compassionate kinship, and a reliance on trickster play and humor. We are the survivors of relocations, dispossessions, and warfare. We are the first people of this land, and my ancestors remind me to remember the regenerative powers of life. "Just like the Thunderers and underwater beings, there are many things beneath and around us to teach us to climb well, to deal with opposition and conflict."[3]

As Seventh Fire descendants, we seek out our original names and clans from elders and teachers and relearn our languages and practices, even if it is sometimes from internet classes, academic books, and state museums. If we are lucky, elders and knowledge holders are available to talk with. In piecing together the ancestral patchwork of the Sixth Fire, I come to understand that even though it is essential to revitalize cultural knowledge and practices and heal from colonial fragmentation, the liberated, uncolonized spirit of the Earth is always available as an ally. In the diversity and beauty of Mother Earth and Father Sky I can find inspiration for my decolonized spirit. Regardless of outer circumstances and projections, my learning spirits can be nourished by the indestructible life force that surrounds us and is within us. As one of my great teachers, the late physicist and philosopher David Bohm, used to say, "There is a reality independent of human thought." Even though we inherit encyclopedias of human thought and collective experience (where much of this is a blessing, a lot of it is limiting baggage) the doorway of the present moment and our interwoven connection to nonhuman life reminds me of a liberated coexistence. "In fact, the waters and rocks can be the very lessons themselves. In an Anishinabek view of the world, the laws of life are all around us." And these natural laws are inherently based on a fabric of mutuality and reciprocity that is ultimately stronger than colonial disruptions.

Given this inheritance and diverse basket of learnings, I want to be an ancestor who nourished legacies of health without forgetting the pain. I want to be remembered as someone who helped create space to nurture a sense of belonging, belonging to that fundamental fabric of wholeness that ultimately created us and will absorb us once again. I want to serve as a hospice worker to encourage the death of colonial, fragmented worldviews and practices that support separation and act as a midwife for a new/old consciousness of justice and harmony, what we Anishinaabeg call the Eighth Fire.[4]

Hospice workers help life die as comfortably and beautifully as possible, embracing the mystery of death as a sacred doorway. There is much in our given society that needs to die: racism, sexism, colonialism, predatory capitalism, greed, or what Leanne Simpson calls "extractivism," the mining of land, waters, human life, for individual profit and greed. Some have called this the "commodification of the sacred."[5] So much of this results from a lack of compassion and empathy and a predatorial, solipsistic existence. There seems to be some obsession and addiction to the power of domination and destruction in capitalistic societies hellbent on Eurocentric notions of "empire," "consumerism," and "progress." Much needs to die and be composted to transform into something truly fertile and nourishing again to all of life, not just for select groups of humans, the one percent. Other related things also need to be ended in terms of ongoing cycles of violence and trauma that perpetuate cycles of neglect, abuse, and violence, and these are related to the structural inequities fueled by the neocolonialism of capitalism, often simply called economic globalization. Ohlone/Esselen writer Deborah Miranda refers to this as the "genealogy of violence."

I decided to end some of these cycles by not having my own biological children. It was perhaps an extreme decision but I felt that I needed to heal myself and make time to understand my inheritance before I could reproduce a healthy human being. I know that children are also very healing and can aid in this process, but for me, it was a matter of focus. Due to my parent's traumas, I had to grow up quickly and often served as a parent for adults, so from a young age I felt that I would choose to be childless. I chose this not only to review my inheritance without perpetuating unconscious patterns onto innocent lives but to lessen the pressure on the finite resources of the Earth, the very "resources," or more accurately, relatives, that held me together as a child. I also knew that there are many parentless children in the world that need care. If I wanted to care for a child, I could always adopt. I also wanted to acknowledge and honor all of the women who worked hard to win women the right to not have children and stand in silent solidarity with all of the women who had children even though it was maybe not their choice or path. On a metaphysical level, I am committed to not being a host for hungry ghosts or a perpetrator of traumas onto new life. I left my children in the unborn, spirit realm. This has been an austere sacrifice but it feels like an authentic part of my life path. It is certainly not for others who enjoy and benefit from the sacred gift of childbirth and biologically producing the next generation, which is a profound and beautiful process. Yet for me, not having my own biological children was a way I could try to end something unhealthy and unconscious and to open space for other non-biological explorations and creations.

This came, of course, with some grieving. Hospice work is often about embracing the grieving process. It includes peering into one's shadow stories and hidden fears. Grieving work involves opening up painful spaces of immense sadness and loss through individual, family, and cultural lines. It can also reveal unconscious shame, guilt, and trauma that can precipitate opportunities for healing and transformation. These opportunities are available in extraordinary and ordinary moments, whenever someone holds a safe space for difficult conversations or for divergent worldviews to collide and coalesce. I want to be an ancestor who facilitated space for these types of inquiries and explorations, to be present and authentic in the face of inevitable pain or conflict. As poet Robinson Jeffers put it, "The calm to look for is the calm at the whirlwind's heart."[6] I do this as a teacher and youth mentor, and most significantly as a lover, partner, daughter, friend, and sister. As the saying goes, when you have no children of your own, all children are yours, and we can serve as nourishing mothers or ancestors at any time as needed or requested.

To balance the hospice work I also want to be an ancestor who served as a midwife and witness to a new/ old consciousness, to the Eighth Fire, the new generation of young people who are strong, beautiful, and united despite pressures to stay isolated and divided. This is the generation of proud Métis, Mestizo, Hapa, Creole, queer, fluid, passionate young people who are healing the relationship between the Indigenous and the settler, the colonizer and colonized, the feminine and the masculine, and are consciously choosing to walk the Green Path, the Pollen Path, the Red Road, and the Beauty Way to a peaceful coexistence on this precious Earth. This generation is rejecting colonial divisions and is rising up and out of the drowning anguish of our times with hip-hop music, Round dance, Earth poetry, and new collective structures (i.e., permaculture coops, artist collectives, and urban farms) dedicated to harmony and justice. They are the Guardians of the Waters, they are Earth Guardians, they are countless young people speaking out about climate disruption and gun violence. Earth Guardians cofounder Xiuhtezcatl Roske-Martinez, who helped file a lawsuit against the US federal government and fossil fuel industry, shares:

> When those in power stand alongside the very industries that threaten the future of my generation instead of standing with the people, it is a reminder that they are not our leaders. The real leaders are the twenty youth standing with me in court to demand justice for my generation and justice for all youth.

Young, strong, and growing, these are the people of the Eighth Fire. You met some of them at the beginning of this section. I have been honored to serve with them as midwife, witness, teacher, mentor, friend, student, and ally.

Together in the San Francisco Bay Area we have been planting seeds, the Three Sisters—Iroquois white corn, with numerous varieties of squash and beans—and other sacred foods given to us by Creator and our web of kin: strawberry, tobacco, sunflower, and others. Through the Cultural Conservancy, we grow these beautiful plants at an organic Farm to nourish ourselves and others and practice food sovereignty. We feast them and learn and teach the whole process from seed germination to various harvesting and cooking methods. Hungry youth come to us seeking to relearn ways to connect to and celebrate healthy, local foods. They come hungry for their ancestral seeds to nourish their bodies and indigeneity. Some also come with handfuls of seeds and their own knowledge of farming and planting, lessons from their families and communities, to share with us. In addition to the delicious cultivated food plants we grow we also learn about elderberry, tule, Oregon grape, manzanita, and willow and other Native perennial plants that grow in Northern California. For the first peoples of this land, these plants serve as the basis for food, medicine, and craft from cough syrup to canoes.

Robin Wall Kimmerer has shared that "to give a gift is to be as a berry in our Anishinaabeg language."[7] I have received many gifts of berries throughout my life: salmon berry, thimbleberry, huckleberry, blueberry, and blackberry from my homelands in Northern California and Juneberry, chokecherry, cranberry, raspberry, and saskatoon berry from my homelands in the Turtle Mountains of North Dakota, to name a few. These berries are my ancestors. Late summers they are usually dark and juicy, sweet and tangy, delectable and intoxicating. Sometimes, though, they are bitter and mealy, dry and acrid. But always they are vivacious in some tangible way. To receive these gifts, to ingest and absorb them into our bodies and beings, is a vital act of nourishment. I have also received such numerous gifts from my ancestors, the sweet and bitter, tasty and sour, all nourishing my body, mind, heart, and spirit in potent and mysterious ways. These berries bind my generations together. So the ancestor I want to be is as a berry—to nourish, nurture, inspire, and vitalize. I can be this as a specific seed planted as a teacher in a classroom with eager youth, and I can also be that seemingly random maple or cattail seed released by the strong North wind to wander and flow wherever it needs to go.

ESSAY

Nourishing

Rowen White

What birth gift should I offer to my grandchildren?
What will help them survive the teetering of the world?
Looking in my basket
I see what it should be—
Selu's heart
The sweet heart of the corn
My children must have whole corn—
The grain and its story.
They will be corn fed like all their mountain ancestors have been.

MARILOU AWIAKTA (CHEROKEE)

Here I am in my happiest place, in the quiet of Dawn, in my cornfield. I begin to prepare the earth for our first of several corn plantings. I hold in my hand a basket of glistening corn seeds, each one a vibrant and deep shade of red, as we are planting our ancestral Mohawk Red corn. This was a variety of corn that I was gifted by an elder when I was only 18 years old, which had been revitalized from just a single cob of corn—an heirloom indigenous seed variety teetering on the edge of extinction from the era of displacement and colonization that my people have endured the last several centuries.

I am always in awe when I hold these beautiful gifts of glistening gems of seeds in my hand as they stir from their sleeping to begin their journey to grow all season to feed and nourish; each one a whole universe of sustenance, their unique seed songs echoing endlessly in such beauty. The early morning sun guides me to the fields, where I truly feel my place in this wondrous universe. This is my sanctuary in this world, in the fields tending the earth; Sun at my back in the eastern sky, raucous raven calling from atop the twin cedars and hummingbirds carrying the day's good news in reverent buzzing songlines. My daughter and I come together this morning in our annual planting ritual. We make a bowl of sacred tobacco and mix it with a little cornmeal from the previous season's crop. We head into the field to the rich earth that has been prepared, and we gently ask permission to come and make our preparations for planting. We sing our seed songs, and offer our handfuls of tobacco and cornmeal to the earth, in an offering of gratitude for the honor of planting here. My toes are warmed by the dark earth, my troubles are carried away on the breeze. Every word is a prayer, every song is an offering, every step is a prostration, every seed the embodiment of hope. Timeless in this moment, where reverence and respect are my clothing and devotion is the sweet nectar that deeply quenches my thirst. In this field, I sit with countless ancestors and hopeful descendants at the grand loom of time, weaving past together with present and future and after a season of hard work and heartfelt prayers, into our harvest baskets comes the gift of a diverse and colorful tapestry of sustenance that will grace the dinner table and feed not only our bodies but our spirits, hopes, and dreams. What an honor and a privilege it is to carry these storied seeds of my ancestors. As my elders say: with privilege comes responsibility. I do my best to keep the seeds alive.

As a Mohawk woman, I say these seeds have been with us since the beginning of time. When I look into this basketful of corn we are planting, I am reminded that Our creation stories never ended, they unfurl into a continuous cycle of creation that inspires renewal each new day.

These seeds are an intergenerational gift, from grandmother to granddaughter since time immemorial. As it has been told to me by my elders, the seeds are a reflection of the people; when the seeds are weak and struggling, it means our communities and nations and people are struggling. When our seeds are strong, it means our nations and communities and people are strong and in good health. These sacred and precious seeds carry our story, sprouting alive into new form to nourish us in many ways. Our beautiful seeds are deeply connected to lineage and specific lands of origin. These foods and seeds are our mirror, our reflection; their life is our life, we are intimately intertwined with their well-being. We are bound in a reciprocal relationship with seeds that extends past beyond living memory . . . these agreements between us and the plant relatives are imprinted into our cellular memory. These food plants and seeds have been with us all since the dawning of our creation stories, they are the extension of a thousand-years-old lineage of responsive and respectful seed stewardship; the seeds always adapting and responding to the changes in the face of our Mother Earth.

In our creation story Original Woman, Skywoman, fell into the watery abyss of this new world clutching seeds in her hand. The sprouting of these ancestral memories was when our foremother, Original Woman, shuffled her feet upon the Earth and sung the seeds and the life into being. When Original Woman's daughter died in childbirth, it was from her grave where her children saw the agricultural food plants sprouting and that was the new dawn of such agreements that we all carry in our blood and bones. The corn sprouted from her breasts, the beans from her hands, the squash vines from her belly button, the sunflowers and potatoes from her legs, the tobacco from her mind, the strawberry from her heart. We still carry one of our original seed songs, that song she sang that inspired life to jump up and live again upon the back of a magnificent turtle, such deep mineral memory that we will sing to our fields each spring.

Years later, in the creation story, Original Woman asks her twin grandsons, Flint and Sapling, the critical questions after they were born, following the death of their mother: "Do you know who you are? Do you know where you come from? Do you know why you are here? Do you know where you go when you leave here?" Sapling answered that he came from the Skyworld, that he came here to make life grow from the Earth and to offer thanks to all around him, and he remembered that he was just a visitor here and he would go back to Skyworld from whence he came when his life was done. Flint, on the other hand, had no idea, no care about where he came from, what he was supposed to do, and didn't care where he was going because he wanted to live forever here. We are reminded in this story to never forget where we come from, why we are here, and where we are going, as we are just visitors here, and someday we will become the ancestors who will turn into the very soil from which the faces of those yet unborn will sprout.

Corn, beans, and squash are the magnificent result of thousands of years of a coevolutionary dance between plants and humans. Corn is not simply a food that has fed and nourished billions over the centuries, it is a representation of our own Indigenous origins, a dynamic reflection of so many unique migrations and cosmogenealogies. Indigenous corn, indispensable deeply storied sustenance of our ancestors, integral in our own cosmologies and creation stories, the seeds themselves become whole memory palaces, helping us remember our lineage, the connection we have to our ancestors, the land, layers of culinary and ecological knowledge encoded into every seed, every benevolent bite of food they offer us. With much reverence and respect, we see corn as our Mother, and from this cultural center, thousands of varieties of corn have been cultivated in the hands of patient and curious Indigenous farmers, corns that have fed humble villagers and built empires. We are kindred relatives, we are bound in a reciprocal relationship with Corn Mother. We are so blessed to continue to learn and live alongside her, and to make an embodied prayer for her continued well-being; as the seed songs fall from my heart and lips like the rain that fell last night, may this last cob to be husked this season be one magnificent living prayer that our children and grandchildren will have plenty of nourishment and food for generations to come. These seeds and foods are our Mothers.

During an era of displacement and acculturation and colonization, some of these ancestral heirloom seed varieties were nearly or completely lost in our communities. Some were carried on long journeys in smoky buckskin pouches, upon the necks of peoples who were forced to relocate from the land of their births, their ancestral grounds. In our work here in this generation, as Mohawk people, we are cultivating the Earth and bringing the seeds back to our cultural center. As one of many Seedkeepers of my community, I hold a responsibility to care for the seeds as I care for our future generations; we do not own these seeds, we borrow them from our children. As a good future ancestor, I feel inspired and grateful to curate a magnificent bundle of seeds to hand down to my children and future descendants. We have beautiful diverse corns, over a dozen varieties of a rainbow of different colors, dozens of different striped and speckled beans, a handful of colorful squashes. We know them as family, as our Three Sisters, each upholding their place in our cosmogenealogy. Each one has unique cultural memories and significance to us; the red corn we need for our wedding ceremonies, the blue corn we need to give our children their names, the white corn is what feeds the people daily, and the multicolored corn is one we use in our children's ceremonies; we have a bean which was said to be the very first bean that sprouted from Mother Earth in our Creation Story. Each one is the reflection of the cultural dimension of biodiversity, as we have coevolved with these plants since beyond living memory.

In our Mohawk stories, our elders tell us our life here on Earth is a gestation, a reproductive cycle that mirrors the life cycles happening all around us. At our birth, we are a sprouting of all our ancestors' wildest dreams, at death we are a sowing of all that we pray and hope for our future generations. While we are alive and walking this pathway of life, we are continually reminded in our ceremonies and stories about the importance of the cultural seeds we can grow and incubate in our lives, in the grand hope that much more life will sprout from our graves and our memory when we become ancestors. It is our obligation as responsible living descendants to do our best to keep our traditions alive, and pass them down to our children.

Now as a mother who sings the sacred seed songs to her children, in a humble act of keeping my ancestral traditions alive, I think of all those who came before me who endured such incredible adversities so that I can stand here today, with seed corn in hand, and love of thousands beating in my heart. I whisper to my daughter:

> Water that seed planted deep inside the earth that is your own body, a tiny seed that sings an achingly beautiful song of remembrance, resistance, resilience, redemption, reconciliation. It was this powerful seed song that kept our grandmothers upright, who whispered to them to get up amidst the sorrow to do what needed to be done to tend the earth and feed the children. It was these melodies that guided our grandfathers under the sea of stars as they made their way into new lands to protect the young. This map is written in the seeds, and the stars and the waters and the Earth. . . . this song is now your heart beating fiercely in promise to uphold the agreements to feed the Sacred hungers of Time.

As a farmer and traditional seedkeeper, I can only imagine and hope that my essence and legacy as an ancestor will be one that nourishes my descendants, in the form of embodied prayers that will reside as seeds that are passed down through the generations from ones that took root here on our family farm. This prayer is one of regeneration, that we will continue in the steps of our ancestors to carry songs, seeds, stories, and prayers that will sustain our future descendants. I can only pray that my memory and my ancestral legacy will be as that of a seed song that is sung from the mouths of my grandchildren who know no hunger.

There is a tiny spirit fire burning in each one of these kernels of corn we are planting under the growing moon; a tiny spark of life that holds the breath and prayer of those who came before us. Those ancestors prayed that we, generations later, would have good food to eat, and clean waters to drink, and good health of mind and vibrant health for all our relations. These prayers now reside within our blood and bones because of the generosity of the seeds, who feed each one of us each and every day. These prayers kindle our own spirit fire to be a continuation of the prayers, embodied prayers on behalf of our children. I feel honored and more alive when I remember that my responsibility and prayer each day is continuing to put my mind and my heart together with all of creation to feed and nourish those beyond this time, because that gift is what has given us life right now. Keeping seeds alive that were passed down through the calloused and loving hands of my ancestors, perhaps even my great-grandmother Anna Jacobs, who loved her garden and to feed her family and community. May our hearts beat in promise to make our lives love poems and honor songs inside of these seeds we tend and care for, so that a little of our life might go forward into the future, in the tiny life capsules of seeds to feed those yet to come. May our daily focus, in whatever small and sacred way, be an honoring song for all the wealth of food and seed and song and story that our ancestors left for us to tend. As I am focused in my daily work in the fields and in my community, I often think of the wise words of Sitting Bull, who said, "Let us put our minds together and see what life we can make for our children."

As stated in a Dakota proverb, "we will be known forever by the tracks we leave behind." I can only pray through our ceremony of everyday that my ancestral tracks will be a life-giving tangle of vigorous green leafy vines and blossoms of all shades, a basket or bundle full of seeds, with my prayers and seed songs vibrating at their core. I think tonight as I shell beans, of my great-grandmother Anna Jacobs, whose legendary gardens and steadfast work ethic are still remembered in our family line. I like to think she may have grown these very beans and that we would have enjoyed shelling them together or trading garden secrets.

What will endure of me when I leave this body of mine, how will I become a good ancestor? It is a life-long love poem that I compose a little more of today in my cornfield.

I make a hole for a little tiny seed. In each little dimple that my finger makes into the warm and moist earth beneath my knees, I place a pearlescent corn seed. I sing a little song, cover it up, and move to the next in the matrix of this cornfield I am planting. I make another hole, another prayer, another seed . . . so on until this whole field is planted. I have my children and family nearby, sometimes we laugh and talk, and other times we plant quietly, softly singing the seed songs that my children have known since before they were born. "Mahji ishka, mahji ishka, manidominess, ashkayna, mino bimadiziwin . . . come in your own time, sacred seed, bless us with life, we humbly ask you that you might please come grant us good life."

As the seeds move into their dark night, courageously they allow themselves to be undone, seed coats cracking and splitting open, and they transform themselves into their imaginal selves into vibrant cobs of corn, multiplying exponentially. These ancestral seeds, who were witnesses to the past, are the keepers of a record of plant-human relationships, one that predates the written word. Within the heart of each seed are countless generations of prayers, ceremonies, and stories, that come alive each season they are planted and harvested and imbibed. If all goes well this season, these seeds will multiply, and will be planted again and again. The seeds will outlive me, but my seed songs will live eternally at the heart of these seeds as they are passed down from generation to generation.

May my life be such a grand sowing, may the seeds I have cultivated and tended to in my life, the speckled ones, the zebra striped ones, the teardrop-shaped ones, multiply exponentially and be shared along kinship routes of the heart to many, perhaps people I will never meet, who will lovingly tend to such earth in hand prayers. May my seed songs live on inside the heart of each of these seeds and plants that I have had the honor to get to know in my meanderings in the dawn; sitting quietly under the waving cornstalks while honeybees gather ochre pollen grains, and the hummingbirds sip dewdrops.

I want to be that ancestor, just like my great-great-grandmother Anna, who didn't forget our responsibility to leave a good bundle of food and seeds and stories for our next generations; who despite all the adversities, the displacements, the busy schedules, the joys and the pains, didn't forget about the seeds and how much they meant to us as people. That amidst the era of the industrialization of our food system, that we didn't forget; we were the ones who remembered to keep the seeds alive for our children; real seeds with flavor and terroir of the land, with the stories and ancestral memories of our ancestors still intact.

Someday my body will return back to the earth from where it came; be the food for the soil people, whose hungry holy mouths eat all the death and decay, the tears and the pain, and create rich renewed earth from which the embodied prayers of my life will unfurl into beans and corn, saplings and mushroom to feed the beautiful bodies and spirits of my descendants whom I will never meet. Until then, I humbly return to the good earth each spring, seed in hand, bowing low to the furrow. . . . . and I plant my prayers that I might be a good ancestor and responsible descendant. That I might be able to make the bundle greater and more rich, colorful and nourishing than when I received it. That I might raise a generation of children who love the seeds and Earth as Mother. My ancestral seed song is a melody that sings:

This is a song of healing through many generations.
A great-great-granddaughter who is allowed to speak her language,
This is the song of a mother who sings the songs of the sacred corn to her children.
This is the song of children being proud of who they are, who they come from.
This is the song of my great-great-grandmother's dreams and wishes coming to life,
In the beat of the water drum and the seeds of the rattle.
This is the song of intergenerational resilience coming alive to dance into another day.

ESSAY

Light

Rachel Wolfgramm and Chellie Spiller

What kind of ancestor do I want to be? How will I be remembered? What traditions do I want to continue? What cycles do I want to break? What new systems do I want to initiate for those yet to be born?

These questions require us to honor their intent and we do this by first locating our stories within the context of "whakapapa" or genealogical narrations from a Māori worldview. The personal narratives that follow are offered by the authors in the spirit of collective learning and we hope they reflect our depth of commitment to the *kaupapa*, purpose, of this book.

WHERE OUR ANCESTOR STORIES BEGIN

We are both Māori and our ancestor stories begin in ancient times captured in genealogical recitals we call "whakapapa." Whakapapa are genealogical narrations that extend beyond the human-centric realms of the secular world as they chronicle and connect the spiritual, social, and natural worlds and the indeterminacy of evolutionary processes. For example, Barlow (1991) articulates a cosmic whakapapa or genealogy of ngā Atua (the gods) that begins with Ranginui (Sky Father) and Papatūānuku (Mother Earth). In this heuristic narrative, complementarity is symbolic, metaphorical, passive, and active as Ranginui and Papatūānuku lay together in an eternal embrace that led to the conception and birth of many children.

The genealogical narrative goes onto tell us that the children of Ranginui (Sky Father) and Papatūānuku (Mother Earth) sought ways of separating their parents because they were discontent with existing in darkness (Te Pō).

Through the act of separating their parents and coming into the world of light and enlightenment (Te Ao Mārama), the children themselves became tutelary gods of the divisions of nature and the environment. For example, Tangaroa became god of the oceans, Tawhirimātea god of winds, Rongo-mā-Tāne god of kumara and cultivated crops, Haumia god of fern root, wild herbs, and berries, Tūmatauenga god of man and war, Rūaumoko, god of earthquakes and volcanoes (Best 1976; Barlow 1991). Whakapapa genealogical-based relationships in creation and evolution are covered extensively by tohunga ahurewa and noted by scholars such as Best, Shirres, Moko-Mead, Hakiwai, Reed, Buck, Hiroa, White, Henare, Taonui.

A major contribution in mātauranga Māori (Māori knowledge systems) was the creation of genealogical recitals or whakapapa to connect humankind to the cosmological and natural realms. Barlow (1991) notes, "When children entered the world of light and dwelt with their mother Papatūānuku, they became ira tangata or mortal beings of the physical world." Tāne-nui-a-Rangi, the first man to inhabit earth, married Hine-ahu-one, the first woman, and begat their daughter Hine-tītama. After many generations their descendants, including Māui, Hema, Tāwhaki, Toi, Ngahue, and Kupe, became famous ancestors of Māori and other islands of Polynesia.

Whakapapa narratives continue to play an important part in the expression of Māori worldviews. Whakapapa is actively used on a daily basis for social and political purposes such as to reinforce iwi, hapū (tribal) and whānau (extended family) relationships, social hierarchies and structures, to reinforce kin-based ties for affective and instrumental purposes.

As active agents, Māori use whakapapa/genealogical narration as a source for renewing spiritual, ancestral, and collective efficacy. How so? Genealogical narrations are transmitted through a variety of mediums, including oral traditions such as those practiced on marae (gathering places) including whaikōrero (oratory), waiata (song), and haka (posture dance). Whakairo (carvings) and tā moko (tattoo) are other mediums that make visible the genealogical recitals. Our buildings, such as wharenui (meeting houses), are physical representations of our ancestors and our ancestors greet us when we walk into them. Although whakapapa belongs to all, and is a way of affirming identity, the creation of whakapapa narratives belonged more exclusively to Tohunga. The process was undertaken in traditional wānanga (places of learning) where the arts, philosophy and native science interacted dynamically in a way that combined intellectual creativity with spiritual meaning and emotional intensity (Marsden 2003). Thus through whakapapa, our relationships to the natural, social, cultural, and cosmological worlds are affirmed.

In outlining a context from which our narratives emerge, we hope we have honored the questions asked. The following section offers two narratives from the authors followed by a brief conclusion in which we offer a summary of our answers.

LIGHT BY RACHEL

My ancestors traveled across vast oceans, inhabiting new homes and places wherever they went. They were explorers, adventurers, warriors, innovators, traders, and settlers. They were themselves "waka aoturoa," driven by curiosity, a search for new environs to explore and settle. Waka is a vessel but the etymology of the word implies the eternal dynamic interaction of light and movement; "wa" refers to time and motion, "ka" to light and fire. Te Ao Tūroa literally means longstanding world, the enduring nature of the world. Use of the terms is often linked to our relationship as stewards and guardians (kaitiaki) of our environments.

My own DNA confirms this story, connecting me to thousands around the Pacific, Native America, Western Europe, and Ireland. Time-travelling alongside my ancestors is enthralling yet what is important to me is how this journey connects us to the present, our environment, and the future. This is an inclusive view, as in my world, my ancestors are all around me, my kin are the mountains, they are the rivers, and the many sentient beings we share this planet with.

In my journeys, I express this connection when I feel sacred energies calling for acknowledgment. I recall a karanga, a vibrational call, I performed on Machu Picchu in Peru, standing beside my sister Tania. This was a sacred moment. My karanga was a greeting to the ancestor mountains of Machu Picchu from my ancestor mountains of Aotearoa. It was a sunny day and as I walked up to the SunGate with my partner, Jim, we heard three distinct claps of thunder rolling off the mountains. I have greeted my ancestors in Hawaii at Hale o Pi'ilani on Maui with a karanga and again, a response arose from the birds, a karanga with the purity of birdsong transfixing me yet transcending time and place. I have looked deep into the eye of a great blue whale and performed a karanga in the oceans of Tonga. A transformational moment, when I saw and felt the wisdom of our collective ancestors through the eyes of our great ancestor, one who can sprout a rainbow when they breathe, who moves elegantly though the oceans in spite of being 200 tonnes, one who can travel the circumference of the moon many times in their life journey, one who can plunge to the vast depths of the oceans and still come up to greet us on the surface. How humble this made me feel, to realize we are so young on this planet we call Home. I wish I could be a wise and humble ancestor too.

This ongoing journey for me, visiting sacred sites around the world, connecting with our ancestors in their natural environments, always reminds us that we are a collective family, one with our environment and with each other.

My inspiration to answering these questions also comes from two of my tribal ancestors. Waimirirangi (sweet waters from heaven) and Hineiahua (the feminine spirit). The names themselves inspire me to articulate our worlds in a way that reflects positively on my tūpuna (ancestors) iwi, hapū (tribes), and whānau (extended family). Both of my ancestors lived in complex times, yet both left incredible legacies for us to follow. These names are embedded in our iwi, hapu, whānau with two of my nieces and many from Northland tribes of New Zealand and the Bay of Plenty tribes of the East Coast gifted with these names.

So to answer the questions, what kind of ancestor do I want to be? I would like to be wise and humble. How will I be remembered? I would like to remembered as curious, an eternal learner, an explorer and an adventurer. What traditions do I want to continue? I want to continue traditions from our ancestors that recognize and value the intelligence of nature, to live with respect and care of our natural environs, and love and care of our whanau. What cycles do I want to break? Cycles that disconnect us from our environment, that are driven by fear, and that lead to violence. What new systems do I want to initiate for those yet to be born? I want my children, grandchildren, and beyond to be part of a movement for good, for peace, for unity and *aroha*.

So my kaupapa, purpose, as "what kind of ancestor do you want to be " is one that is dedicated to the pursuit of these things and that leaves ancestral footprints like those of my own ancestors that tell a story of curiosity, exploration, and adventure, but done in the spirit of humility, unity, peace, wisdom and aroha.

Ki a tipu ka rongo i te ihi, ihi o te ra, ka rongo i te ra, ke whakamahana i te ara . . .

"Every living thing has a pathway of growth and life, listen to the voice of the dawn maiden heralding the coming of the sun, feel warm rains of spring as Hinekohu puts a korowai over her mother . . . ka ihi o te ra . . . ka haumai to au e rere nei . . ." (Fraser Puroku Tawhai, Te Whakatohea)

LIGHT BY CHELLIE

In 2016 I went on a journey into a forest deep in the heart of Aotearoa, New Zealand, with my Native Hawaiian friend Dr. Elizabeth Lindsey and another friend, "D." Elizabeth and I were writing together, exploring our shared love of wayfinding and ancient cosmologies. I had written a book, *Wayfinding Leadership* (2015), and Elizabeth had been a student of one of the greatest celestial navigators in the world, Mau Piailug, for 10 years while completing her PhD in anthropology with a specialization in ethnonavigation. Before Mau passed away he shared with Elizabeth what he called "the code" to wayfinding, "It's all to do with love, light, and frequency," he said. After years of research, travel, and inquiry I had a direct experience of Mau's "code."

The three of us stood at the entrance to the ngāhere, the forest, Elizabeth, D, and I. We held hands and I did a karanga, a vibrational call, to acknowledge and show respect for the sacred domain we were about to enter.

We moved through silently to the place where the day before we had cleansed the taonga, the precious objects we had brought with us for cleansing. Lying on the mossy bank at the spot where we had crouched down to place our taonga in the sacred waters was the pounamu, greenstone, that had had gone missing, which was why we were back again. To look for it.

Where the day before the lake had been a sparkling aquamarine, so clear that we could see a number of trout suspended in its pure, deep waters, today a wind rippled its surface into an opacity upon which sunlight blinked.

D took off her shoes and found a spot to sit and put her feet in the lake. I sat nearby and plunged my feet into the cold water. My feet ached with the cold and I told myself to keep them in the water until they felt warm. As we sat there after some time we heard a chant coming across the other side; it was Elizabeth intoning an ancient Hawaiian prayer. It traveled across softly with the wind. I felt a touch on my right shoulder, it was D. She was pointing to a spot of light not far from where she was sitting. It was moving towards us.

At first, I didn't know what to look at, what I was meant to see. All I could see was a tranche of light dancing on the surface of the water. Then, some kind of awareness settled in me. I looked without trying so hard to see. I became an awareness of being with the light. The most amazing thing happened. Instead of just being an undifferentiated spot of light it began to separate out into star bursts, quite distinct from each other with space in between each. Each starburst was like the most gorgeous and beautiful snowflakes except these were of light. Light crystals. The light crystals were blinking. I just stayed in a state of awareness and presence until the light crystals become just three or four and then were none.

It had probably lasted only about 30 seconds, I don't know how long D had been touching my shoulder, I was aware of energy running through me. I was aware that I was fully present with no clinging to the experience, no wanting the light crystals to stay. It felt like joy and love. Of being communicated with through frequency—a signaling whose code I cannot read. It seems I had just been given a direct experience of love, light, and frequency.

Sitting here and recounting the experience I see that no words, no readings, no theory, no discussion could ever provide me with the knowledge that I received through that direct experience. A profound beauty in the world had been revealed, and I had peered into a deeper and exquisite fold of reality. Did Mau read the signals of love, light, and frequency as you and I would read a book? Were they communicating with him—showing him the way? Will I ever see like that again? Water crystals, snowflakes of light. It's a beautiful mystery.

The closest image I could find was in the mystical scientific (and some say controversial) work of Dr. Emoto, who experimented with photographing water molecules—the one associated with Love seemed closest, although what I saw was different. Each star burst had distinct 'arms' with a circle at its tip.

Not long after this experience I was rereading Annie Dillard's book *Pilgrim at Tinker Creek* and came across this wondrous passage that resonated so strongly with my experience in the forest:

When her doctor took her bandages off and led her into the garden, the girl who was no longer blind saw "the tree with the lights in it." It was for this tree I searched through the peach orchards of summer, in the forests of fall and down winter and spring for years. Then one day I was walking along Tinker creek and thinking of nothing at all and I saw the tree with the lights in it. I saw the backyard cedar where the mourning doves roost charged and transfigured, each cell buzzing with flame. I stood on the grass with the lights in it, grass that was wholly fire, utterly focused and utterly dreamed. It was less like seeing than like being seen for the first time, knocked breathless by a powerful glance. The flood of fire abated, but I'm still spending the power. Gradually the lights went out in the cedar, the colors died, the cells un-flamed and disappeared. I was still ringing. I had been my whole life a bell and never knew it until at that moment I was lifted and struck. I have since only very rarely seen the tree with the lights in it. The vision comes and goes, mostly goes, but I live for it, for the moment the mountains open and a new light roars in spate through the crack, and the mountains slam.

I hope that many generations from now people will still be "knocked breathless" by the lights in the trees and the lakes, that they too are "spending the power" in a world still full of mysteries, sacred places, and kaitiaki, guardians. To answer the question what kind of ancestor would I like to be? I would like to be the kind of ancestor that I look to in my own ancestors; when I need a korowai, a cloak, of strength, courage, and love to guide me. Perhaps in some sense I will be a form of love, light, and frequency that comes as a presence.

How will I be remembered? A long time ago my elders and mentors called me to be a Ruahine, someone who moves between worlds, to work between worlds, the spaces in between, to be comfortable with both the known and the unknown, the seen and the unseen. The spiritual and the material. I'd like to be remembered as a Ruahine explorer who went on inner and outer wayfinding journeys of discovery.

What traditions do I want to continue? A deep belief in a sacred ecology, that we share kinship with all of creation and it is our responsibility, as kaitiaki, guardians to look after our planet.

What cycles do I want to break? The notion that we are separate, individualized individuals who can produce and consume what we like without regard to the impact we are having.

What new systems do I want to initiate for those yet to be born? Our ancestors bequeathed a knowledge system for us to live by, a system built upon principles of reciprocity and respect. The wisdom contained in Māori values, developed over the aeons in relationship to the world around us, help us be kaitiaki of our planet and each other. Values such as whanaungatanga, creating community and cultures of inclusion; manaakitanga, showing kindness and respect to others through manaaki, uplifting the mana and well-being of others; kaitiakitanga, being a kaitiaki, steward and guardian that nurtures the life-sustaining capacity in the world; and hūmārietanga (humility) where there is more quiet, more space to see what is going on around us, to hear others, to tune in. It is the art of listening. It is the space we create.

SELECTED BIBLIOGRAPHY

Barlow, C. *Tikanga Whakaaro: Key Concepts in Māori Culture*. Auckland: Oxford University Press, 1991.

Best, Elsdon. "Māori Religion and Mythology Being: An Account of the Cosmogony, Anthropogeny, Religious Beliefs and Rites, Magic and Folk Lore of the Māori Folk of New Zealand, Part 2." *Dominion Museum Bulletin* no. 11 (1976). Wellington, Museum of New Zealand, Te Papa Tongarewa.

Dillard, A. "Seeing." In *Pilgrim at Tinker Creek*. New York: HarperCollins e-books, 2009.

Emoto, M. *The Hidden Messages in Water*. Hillsboro, OR: Beyond Words Publishing, 2004.

Henare, M. Tapu, Mauri Mana, and Wairua Hau. "A Māori Philosophy of Vitalism and Cosmos." In *Indigenous Traditions and Ecology, The Interbeing of Cosmology and Community*, edited by J. A. Grimm, 197–221. Cambridge, MA: Harvard University Press for the Centre of the Study of World Religions, Harvard Divinity School, 2001.

Howe, K. R. *The Quest for Origins: Who First Discovered and Settled New Zealand and the Pacific Islands.* Auckland: Penguin, 2003.

Marsden, M. "Man, God and Universe: A Māori View." In *Te Ao Hurihuri: The World Moves on Aspects of Māoritanga*, edited by Michael King. Wellington: Hicks Smith, 1975.

Marsden, M. *The Woven Universe: Selected Readings of Rev. Māori Marsden*, edited by Te Ahukaamaru Charles Royal. Otaki, NZ: Estate of Rev. Māori Marsden, 2003.

Moko-Mead, Hirini. *Te Toi Whakairo: The Art of Māori Carving.* Auckland: Reed, 1986.

Moko-Mead, Hirini. *Tikanga Māori: Living by Māori Values.* Wellington: Huia, 2003.

Salmond, A. "Tipuna-Ancestors: Aspects of Cognatic Descent." In *A Man and a Half: Essays in Pacific Anthropology and Ethnobiology in Honour of Ralph Bulmer*, edited by Andrew Pawley, 334–47. Memoir no. 48. Auckland: Polynesian Society, 1991.

Spiller, C., H. Kerr, and J. Panoho. *Wayfinding and Leadership.* Wellington: Huia, 2015.

Te Awekotuku, N. "Māori People and Culture." In *Māori Art and Culture*, edited by Doro Starzecka. Auckland: Bateman, in association with British Museum Press, 1996.

Walker, R. "The Relevance of Māori Myth and Tradition." In *Tihe Mauri Ora: Aspects of Māoritanga*, edited by M. King. Wellington: Methuen New Zealand, 1978.

Wolfgramm, R., and E. Henry. "Ancient Wisdom in a Knowledge Economy before and beyond Sustainability: Valuing an Indigenous Māori Perspective." Conference Proceedings, Inaugural International Conference for Indigenous Enterprise, Albuquerque, NM, 2006.

INTERVIEW

Ilarion Merculieff

Brooke Parry Hecht

HECHT: *In your book,* Wisdom Keeper, *you wrote of how your Elders answered questions in a manner that was responsive to the questions that were asked. You said that the answers you were given "came from their assessment of how much I understood, as reflected in the question I was asking."*[1] *So, with my deep gratitude for opening your heart to my questions, I'm also asking for your help in shaping the questions themselves, so that we can get to the heart of what needs to be shared at this time.*

MERCULIEFF: Okay. Sure.

HECHT: *Could you share the earliest stories you carry of your ancestors? I'm interested in knowing how the Unangan came to be on the Aleutian Islands and how they believed one should live.*

MERCULIEFF: We have stories that take us all the way back to Egypt. We traveled from Egypt, to Mongolia, out of Mongolia and Siberia then across the water by kayak and large craft to Kamchatka. Those are our origins. We were a very spiritually based culture. For example we were the only people in Alaska that didn't have footwear, even in the wintertime.

HECHT: *Wow.*

MERCULIEFF: That's something very few people know, but it's true. Even today, when we do performances with our regalia, it's done without any footwear.[2]

HECHT: *Could you say more about how the footwear relates to spirituality? In other words, what I'm imagining is that the ground is cold and covered with snow, but your feet are warm.*

MERCULIEFF: Yes, that's right. It's very much like the stories you have heard of the Lamas that sit out on snow and melt the snow. A forensic anthropologist who studied Unangan history and bones once told me the average Unangan (Aleut) man was a world-class athlete—literally, that every one of them was like that. He took my biceps and said, "Make a muscle." So I did. Then he took his hand and doubled it and said, you know, "The average man was like that." At age five, to be initiated, they had to sit out in the Bering Sea for four hours a day, for four years—even in the wintertime. Every single day they would sit out naked in the water. And then right after that, they would run up and down the hills with boulders in their outstretched arms. This was to prepare them to use our high seas kayaks. Our people were known to have taken high seas kayaks all the way to the end of South America.

HECHT: *I loved your recounting of that South American connection in your book. When you visited the Mapuche Elders in southern Argentina, they were so thrilled to meet you—and experience for themselves the kinship and connection that has existed between the Unangan and Mapuche, going back thousands of years.*

MERCULIEFF: That is so true, and I'll never forget that visit. The single-holed kayak the Unangan used is called an *iqyax*, and the two-holed kayak is called an *ulooxtahn*. I built an *iqyax*; it took me nine months to make. It was appropriate that it was nine months, the gestation period for a human being. In making that craft, I began to see and understand how our people adapted and how they literally became one with the environment. So, as I said of the people who would go without footwear, the same principle applied to the kayak. In the water, we become one with the environment. Our crafts were the first kayaks made with ball bearings in them. The ball bearings are made out of ivory at every joint in the boat, and when you're out in the water, the boat moves with every single nuance of the water. So you literally feel that you are just part of the water. It is quite, quite amazing.

We also had a basic form of brain surgery. This was all pre-written history, before the Russians arrived. We even mummified our dead, in a very similar way to the Egyptians. The only difference being the place where we cut a hole in which the insides of the person were taken out. Our hole was on the side and the Egyptians' hole was in the middle. What made our practice unique is that we were able to mummify our dead in such a moist climate. The Egyptians, of course, had a very dry climate.

The Egyptians and the Unangan were the first two peoples in the world that had a word for "billion." And that's before any Westerners arrived to our island.

HECHT: *Do you know what the word was used for, or why you had a word for "billion"?*

MERCULIEFF: [Laughs] Yes. And I use that when I do my talks, and I have people guess.

HECHT: *Is it for birds?*

MERCULIEFF: No, you can't count a billion birds. You can't count a billion anything, for that matter. Even the stars, you can count maybe two thousand. Why would our people have a word for "billion"?

HECHT: *I'm going to have to think about that one and let you know if I think of something.*

MERCULIEFF: Okay. Yeah.

HECHT: *You've said that your ancestors came from Egypt and then through Mongolia and Siberia. Do you carry a story of why your ancestors left Egypt or why they stopped at the Aleutian Islands as they did? Why did this become their home?*

MERCULIEFF: Why they stopped in the Aleutian chain? That I don't know. When you look at the Aleutian chain, it seems pretty bleak. There are no trees, there are lots of mountains, and it's basically sub-alpine tundra. So, most people would think it's bleak. How do you eke out a living in that kind of environment? But again, as a testament to the ingenuity of my people, we had the densest population per linear mile shoreline than any place in North America. And our people never had any intentional food storage technology. As one of my Elders on St. Paul Island said, even the birds don't worry about where they're going to get their food the next day. That's the principle on which we operated. We were so in touch with what I would call a Divine—every day, every minute of every day. We trusted in life, in the life processes of the universe, in the role of the Maker, that we will get what we need. The reason we were that way is that we were present in the moment and in the heart, what we call the real human being. The real human being is someone who has self-authority that is shared with everybody in the group, someone who lives each day as a new life, basically. And so, we had a very spiritually based culture for our time.

Of course, much of that has changed now. The Russians arrived in 1741, and they decimated our population. Within fifty years, about 80 percent of our population died due to malnutrition, starvation, disease, and genocide. Eight out of ten people were gone. The first people the Russians took, or killed, were the shamans and their apprentices. And then they went on from there. From the survivors of our holocaust, they handpicked the best hunters and their families and moved them to the Pribilof Islands. We call it Tanax Amiq, which means "land of our mother's brother." We always knew about the existence of these islands and kept it secret from the Russians. But the Russians heard the stories and decided to look for these islands, and they found them. And when they found them, they handpicked the men and their families, and went to these islands.

But the reason the islands had never been occupied was that long before the Russians settled us there, Igadagik, the son of a chief, was on a hunting trip on a kayak, and he got blown off course, north of his home, Unimak Island. He got lost in the fog until he heard the bellowing sounds of seals. Igadagik could hear them in the fog. When he went toward their voices, he discovered an island full of fur seals. Under the Russians, that island became St. George Island. When Igadagik went on the island, he went to the highest point on the island. From there, he saw what the Russians would call St. Paul Island. At the time of Igadagik, it was estimated that there were four million northern fur seals there. One day, on a clear day—Igadagik was there for about a year—he looked and he saw the tops of the mountains of the Aleutian chain, and then knew the direction he needed to head out to go back to his islands. He told the people when he got back to his village what he had found, and the shaman said, "Okay, we'll go there again, and we'll see what it's about." So, a hunting party and the shaman went to the island, and as soon as the boats landed, the shaman knew what this was. And he told the people this was not a place to live. Because it is—I'm sure he didn't use this word at the time—but it's a power point, a power place. You can visit it, you can have hunting camps there, but you cannot live here. And so that's what our people did, until the Russians took us to the Pribilof Islands in 1786 and 1787.

There, we were slaves—the survivors were slaves of the Russians. The Russians held control over us until the jurisdiction of Alaska was sold to the US. We say jurisdiction of Alaska—Westerners say they sold Alaska. But when you look at history, you will know that Alaska was never sold.

Under US jurisdiction in 1867, we became slaves for the US government. We didn't achieve political independence until 1966.

HECHT: *When was Igadagik's kayak blown off course? Is this an old story or more recent history?*

MERCULIEFF: The best we can tell is sometime between 4,000 and 8,000 years ago.

HECHT: *So it's an ancient story. And from that point forward, the shamans held that, forever, this is a sacred place and, as you said, a power place. Of the Unangan who survived the Russian genocide, the strongest hunters were taken and made to move to this power place—where the Unangan knew that humans were not supposed to live—and then Russians forced them to harvest seals there?*

MERCULIEFF: Correct.

HECHT: *And Unangan people still live there today?*

MERCULIEFF: That's right, and of course we know the environmental effects that are occurring there, and the overfishing from industrialized fishing is fundamentally changing the Bering Sea's ecosystem today. So when Igadagik found the island, there were four million fur seals, and today there are 400,000. When I left there to live in Anchorage thirty years ago, there were 750 people and about 1.2 million fur seals. So seal numbers are going down, and going down fast, because females are disappearing. We know that they are disappearing because of lack of food, and they are experiencing food stress.

Two years ago, we had the first ever recorded reproductive failure of 100% of all sea birds on the Pribilof Islands. Before, we would have one or two species that might experience reproductive failure, but never all of them. And, we know that it's an ecosystem-wide phenomenon because near-shore foragers, distance foragers, depth foragers, surface foragers that all live on the island are experiencing real trouble. So, we've noted, for example, sea lions chasing after fur seals in greater frequency than in living memory and eating them a lot more voraciously. Chicks have been falling off cliff ledges in large numbers—and it wasn't an isolated phenomenon; this happened year after year after year because they were too weak to maintain their hold of the cliff ledges. We saw the adult birds with their breast bones sticking out and chest muscles caved in. When you took the flesh off fur seal pelts, you could see light through them; we'd never known that before. So, we knew that there was an ecosystem-wide phenomenon.

We pointed this out to officials that were in charge of fur sales back in 1977—and, ever since then. Of course, they took our knowledge as anecdotal information, bits and pieces of information that may or may not be useful to Western science. Of course, had they listened to us, they would have saved years of trying to do research on what was causing the decline of so many species. But, live and learn.

HECHT: *I have a question about the nature of the power place. What does it mean that there are people living there now? We know their ancestors were brought there. They were forced to live there, and their descendants are there now. I don't know exactly how to phrase my question, but this might be close: Are the shamans able to work with the power place now? Can the sacredness of a place change such that it is okay for the people to live there?*

MERCULIEFF: First of all, I don't think there is anything that can be done to a place that makes it okay to live there if, spiritually, it was decided that we shouldn't live there. That's shown by what's happening to the island—the animals are declining, the people are declining. When I left there, there was a population of 750 people; now it's 450. So, it's down by three hundred since I was there thirty years ago. And, the people suffer emotionally. They have trauma that is intergenerational. It affects just about everybody on the island. We never had any suicides until the government pulled out of the Pribilof Islands in 1983. In 1983, we had one hundred documented suicide attempts in a population of, at that time, about 650. And, we had four murders. By contrast, in the prior one hundred years, we had no suicides, and one murder. So, the situation got very grim. And they are trying to adjust themselves now. The tribe has essentially taken over. The city has gone down to one-tenth of what it used to be, and the tribe is trying to do what it can to protect the environment around the island of St. Paul. St. George is applying for designation by the federal government that would increase the protection of the waters around the island. Of course, we don't know what that's going to be because there's lots of opposition to it by industrial developers and fishermen. So, they are trying to correct the wrongs that have been done out there, but it's anybody's guess as to whether or not we'll succeed.

HECHT: *It's so hard to imagine how such transgressions and such deep wounds can be healed. You tell a story about your experience facing a historical trauma on an island where there was a terrible massacre. In* Wisdom Keeper *you recount the history of the Russian fur traders killing and enslaving your ancestors in the 1760s and how, at that time, some people escaped to a small island off Unalaska Island.*

MERCULIEFF: Yes, Unalaska is the biggest island there in the Aleutians. And the story is that Soliev, a Russian captain, wanted revenge for an uprising that our people had carried out in the early 1760s. By 1768, he organized the Russians to kill every Unangan person living on the three largest islands in the Aleutians: Umnak Island, Unalaska Island, and Akutan Island. He wiped out a lot of our people during that time, but the survivors—mostly old people and children—were able to get to an island that could be comparable to Masada. It was a place we tried to have the chance of survival. The island is off of Unalaska Island. I went to this island with the Discovery Channel in the eighties. We wanted to verify that these stories that our people carried are true about this Unangan Masada. None of our people had gone back to that island since the Russians discovered it—discovered our people there—and wiped them all out in four hours. And, there were about four thousand Unangan people that were killed in four hours. And so we wanted to see how accurate our people's stories were.

So, we went to the island, and the Discovery Channel created a film out of this called "The Mystery of the Alaskan Mummies." I was there as the Unangan representative to make sure that they didn't do anything that would violate this area. One of the things I asked them to do was have a tight shot of the island so that you couldn't see the shape of the island, for fear that hunters of artifacts would raid that island. Anyway, we got there, and we did verify that there were ulaxs there—semi-subterranean houses—and that the estimate of four thousand people was roughly correct, based on the number of occupants that could be inside this structure. We verified that the Russians had dumped kegs of gunpowder down a hole on the top and blew up everybody inside. And there were thousands of musket balls and hundreds of cannon balls all over the place. We found one Unangan person who was—we had a forensic anthropologist with us from the Smithsonian—and we found a skeleton. What we determined was he was a seventeen-year-old boy, and he was killed. The knife was still sticking in his chest—the bones of his chest. And that he was probably dragged there and covered with a scapula, whale scapula, in a hurried state, because that's not the way we bury our people. And so I gave them permission to take the knife as a reminder of our holocaust, because our people have never grieved our holocaust before, ever. They haven't, still. And so, this was a reminder that we still have to grieve. And so that's now in the museum in Unalaska.

HECHT: *This is so painful to hear. It's just very painful. You wrote that when you arrived with the film crew at the island, there were two eagles, and that these eagles had been showing you . . .*

MERCULIEFF: I didn't know I put that in the book, but yeah!

HECHT: *You said they'd been guiding you and that when you arrived to the island, they drew you to a certain place where you felt compelled to sing and drum. And that then your ancestors spoke to you clearly.*

MERCULIEFF: Yes, they did. Now, what I didn't say, I don't think, in that book was that as we approached the island, the director said, "We're going to move over here." So I looked for the two eagles and didn't see them at all. So I knew that that wasn't the place that we were supposed to be, so I told the director. He said, "Well, yeah, okay, but you know, we can walk across the island if we have to, it's small enough." So, we went and landed, and as we landed there were spirals of seagulls up in the air right above us, and then all of a sudden one of the seagulls and then another and another were basically shitting on us . . . And the only person that wasn't hit by the guano from these birds was me!

HECHT: *And this was where the director had wanted to land. And you were telling him, "It's the wrong spot!"*

MERCULIEFF: Yep. And so, he looks at me and says, "Okay, we'll go wherever you say we should go." And so, we went back out, and I looked for the two birds and saw them, and so I went to this one place. That was exactly where the tragedy occurred. And then I went to the edge of the island while the other people were at the site. I went to the edge of the island and I did sing and, the thing was, my drum didn't make a sound. It was the island that was making the sound; it was the island that was drumming. When the film crew came looking for me, they saw me at the edge of this cliff on the opposite side of where they were, and they heard the island drumming. And so, they figured, "Well, we better get out of here."

HECHT: *And what did you think?*

MERCULIEFF: Well, I knew that this was a very spiritual experience, and there has to be a reason for this. And during that time, after I stopped drumming and seeing, I heard this distinct voice, that they were my ancestors, and they're still there, and they're going to be there until the people grieve what happened to them as a whole, not just what happened on that island. That was the message I got.

HECHT: *Wow. So that was the message. I want to relate this story to another story you tell about receiving important spiritual guidance, in this case about oppressors, those committing genocide. There, the message was, "They do these things because they are asleep in Spirit. It is a spiritual sickness from which they suffer, and they do not know or understand this. And they can't know any better way, unless they're shown a better way. A way of love, a way of compassion, a way of wisdom." You were at a gathering of First Peoples, grappling with this rift between some First Peoples and Westerners, and that was what you were guided to say to the First Peoples present.*

For the Unangan, your name means bridge—like an arm coming out to the world from the Unangan. Do you have a message for those drawn to the other edge of your bridge, whether oppressors or the descendants of oppressors? Is there a message for them?

MERCULIEFF: Yes. I'm starting to see this now. I've traveled all over the world, and I just came back from Israel about three months ago. What I'm seeing everywhere I go is that people are starting to wake up. They are at the point where they're saying, "Okay, we can see that nothing's working now that used to work. Environmentally, politically, socially, political injustices, refugees, the wars . . ." I mean it's all coming to a head. They're seeing that nothing is working that they have tried. And for example in Israel of course, they have been praying for peace for generations. But, there doesn't seem to be an answer to that prayer. And they're at the point where they're saying, "Okay, what do we do now, if our current work is not doing anything?" And in fact, things are getting worse. When you look at the environmental organizations, there are thousands more environmental organizations today than there were thirty years ago. But yet, the life support systems of Mother Earth are being pushed to the edge.

Why is that? Well, people are starting to see what is happening, and they realize that the brain or logic centers—the reasoning centers—are not the place for answers. If we go to the mind as the center of intelligence, it's the wrong place, because the center of intelligence is the make-up of the total human being, which includes intuition, heart-sense, gut-feel, all of these things which are discounted by Western society. And so, I tell people what the Elders say that I work with. They say that the most unselfish thing you can do to help in the world today is to heal yourself. And so what that means is changing one's consciousness from the mind to the heart. Because the heart is the place of the connection to the Divine. It's the heart that tells us what to do. The mind's job is to figure out how to do what the heart is telling us. But we've reversed that today so that the mind tells the heart what to do.

That's the message that we're carrying through a meeting of world Elders that we had in November of last year in Kauai, Hawaii, which is the birthplace of the souls. We recorded thirty hours of film footage. We asked the Elders what we could film and what we couldn't film. And the response was unprecedented. They said, "We don't want to filter any of this. You can take films of all the ceremonies that we conduct—and all our discussions." That's unprecedented for Native groups in the world. Because the Elders said, "The time for secrets is done. The time of lone wolf is done. We must act now to change our consciousness."[3]

HECHT: *That is amazing. I have a follow-up question regarding oppressors and descendants of oppressors. Do you have insights to offer for those who are cut off—by many generations—from practices in relationship with the homelands of their ancestors? In your chapter in Wisdom Keeper, you say the link between "older experienced hunters and the young men would be severed if the ability to hunt and fish for traditional foods is lost. The link between the older women and younger is severed if the ability to gather wild foods is lost. It's in the 'subsistence camp' that the younger have extended contact with the older to learn language in the actual context of real-life experience from which the language was born, the ethics and values of being a real human being, such as how to cooperate in a group, how to share foods with others, how to respect and have reverence for fish and wildlife, how to demonstrate respect for Elders, the lands, the fish, wildlife, and the waters."[4] I'm guessing that those of us who have been cut off from such practices for generations don't even know the problems that we are carrying as we try to navigate the cultural waters we live in now. So, here we are now. Here I am now. How can we honor the practices that are of this place and are original to this place without appropriating them?*

MERCULIEFF: I was up in Canada, in Alberta, visiting some friends, and I got invited by the Elders at the reservation in Morely, a four-thousand-acre reserve, to be with them when they were going to conduct a ceremony that they hadn't performed, even in secret, for a hundred fifty years. I said yes,

I'll go. There was nothing but white hair, they were all white-haired, and they were wearing buckskin, and it was at the battered women's shelter. And they had four sacred pipes in the four directions. And the ceremony lasted for four hours and they only spoke their own language. They didn't speak any English until the spokesperson, about two hours into the ceremony said, "We're speaking English for the benefit of our friend from Alaska. We know why you're here." And of course, I thought I was there because they had invited me, but they knew differently. They said, "We have been praying to the Creator, and we have a message for you to carry, if you will. The message is: We know that there are many people who think that they have forgotten their spiritual ways, and that it's never going to be returned again. We want you to know that this is not so. It has been kept for you in the unseen world, waiting for you to wake up in Spirit." And that was the end of his message, and they continued on with their ceremony. Well, that is so true, that when we wake up in Spirit the things that we thought we lost are returned to us. It's slow, but it depends on where the person is. I know this because I went through that. What it means—waking up in Spirit—is you must get in touch with your heart. Your heart will guide you as to what you need to do—you yourself. Each person is unique, and each person has their own heart response. And a heart response is what is needed today, now. There is no more time to dawdle in the mind. We've got to become whole again, and in intimate connection with the Divine, and the only place that we can do this, the only place, is the heart. And the heart will never guide you wrong. It comes out with things like: This is a situation that requires compassion. This is a situation that requires love. You've got to love yourself. You've got to break free of all the traps that you've learned since the beginning of time—and time began when we left the present moment. It was human beings that created time. And, the only place that we can find our connection to who we are is now here in the heart present in the moment. And that's what the Elders are saying in the messages.

HECHT: *Thank you so much for this, Ilarion. I will close with one more question. The first time I met you, my daughter Tara was eight or nine. And I told you that she had just "discovered" that magic wasn't real. I told you how she was feeling shattered by this. And so on the inside of your book, Wisdom Keeper, you wrote to us, "Dear Brooke and Tara, I hope you enjoy these stories. Our lives are filled with magic, and many adults don't remember this, but it exists. It took twenty years to write this story filled with magic,* even though there were hard times. *Never forget that it exists, no matter how old you are."* Is there anything you want to share to remind young and old, anyone reading this book, about magic and, perhaps, the re-enchantment of our world?

MERCULIEFF: That's what keeps us from seeing what is out there. I mean there is magic in life every single day if you just have the eyes and ears and senses to see it, and to feel it. And, as children we all carry this understanding of magic. At some point then something happens and we are told not to believe in it. And belief is not something that is a word that I use with myself. I don't believe, I know from my own experience. So, I know things but each person is going to know something different. And each person is on a sacred path, and we must recognize those sacred paths, and learn them. As children, though, we all have this understanding of magic, and we somehow lose that. But it's those people that hold onto that childhood knowledge into adulthood that become the spiritual people. And the magic doesn't go away. It's still there. And it's there within you and outside of us. And we can recover it at any time, no matter what age you are.

HECHT: *Thank you so much for sharing the wisdom you hold. It is a gift to all of us.*

POEM

Lost in the Milky Way

Linda Hogan

Some of us are like trees that grow with a spiral grain
as if prepared for the path of the spirit's journey
to the world of all souls.

It is not an easy path.
A dog stands at the opening constellation
before you can reach the great helping hand.

The dog wants to know,
did you ever harm an animal, hurt any creature
or did you take a life you didn't eat?

This is only the first of the map. There is another
my people made of what is farther
beyond this galaxy.

It is a world that can't be imagined by usual means.
After the first,
it could be a map of forever.

It could be a cartography
shining only at some times of the year
like a great web of finery

some spider pulled from herself
to help you recall your true following,
your first breath in the dark cold.

The next door opens and Old Woman
counts your scars. She is interested in how you have been
hurt and not in anything akin to sin.

From between stars are the words we now refuse;
loneliness, longing, whatever suffering
might follow your life into the sky.

Once those are gone, the life you had
against your own will, the hope, even the prayers
take you one more bend around that river of sky.

ACKNOWLEDGMENTS

Our deepest thanks go to the ancestors, from those we might be lucky enough to know of from stories to our most pervasive ancestor—the land and its communities that sustain life itself. We have been humbled by the compelling and creative contributions of the authors of this book, always and already ancestors that regenerate mind and heart. This book would not have been possible without the platforms and organizations that brought us together—Black Mountain Circle's Geography of Hope, the Center for Humans and Nature, Western Colorado University's Headwaters Conference, and the Cultural Conservancy. Finally, we want to acknowledge the human and more-than-human ancestors that will emerge from the lives of our generation's descendants. If our "Seventh Fire" part is any indicator, we find great hope from the critical wisdom of those leaders who are already emerging and who are to come.

NOTES

NOTES TO PART I

Grounded

1. L. E. Graham, M. E. Cook, and J. S. Busse, "The Origin of Plants: Body Plan Changes Contributing to a Major Evolutionary Radiation," *Proceedings of the National Academy of Sciences* 97, no. 9 (2000): 4535–40. doi:10.1073/pnas.97.9.4535

2. E. Salmón, "I Want the Earth to Know Me as a Friend." Questions for a Resilient Future: What kind of ancestor do you want to be? Retrieved August 21, 2017, from https://www.humansandnature.org/i-want-the-earth-to-know-me-as-a-friend

3. W. LaDuke, "How to Be Better Ancestors." Questions for a Resilient Future: What kind of ancestor do you want to be? Retrieved August 21, 2017, from https://www.humansandnature.org/how-to-be-better-ancestors

4. K. Whyte, "Settler Colonialism, Ecology, and Environmental Injustice," *Environment and Society* 9, no. 1 (2018): 125–14. doi:10.3167/ares.2018.090109.

My Home/It's Called the Darkest Wild

1. Cold Mountain, *The Collected Songs of Cold Mountain*, trans. Red Pine (Port Townsend, WA: Copper Canyon Press, 2000).

NOTES TO PART II

Of Land and Legacy

1. E. Vallianatos, September 10, 2012. America: Becoming a Land without Farmers. Retrieved July 17, 2018, from https://www.independentsciencenews.org/environment/america-becoming-a-land-without-farmers/.

2. R. Grant Seals, R. C. Wimberley, and L. V. Morris, *Disparity* (New York: Vantage Press, 1998).

3. Jeffrey Sachs, *The Age of Sustainable Development* (New York: Columbia University Press, 2015).

Cheddar Man

1. Arnold Mindell, *Dreaming while Awake* (Charlottesville: Hampton Roads, 2000).
2. I find it easy to imagine that these early hunters had a language and therefore names for one another. And who's to say that one day scientists won't be able to extract this information from Cheddar Man's bones.
3. Robin McKie, "Cheddar Man Changes the Way We Think about Our Ancestors," *Guardian*, February 10, 2018, https://www.theguardian.com/science/2018/feb/10/cheddar-man-changed-way-we-think-about-ancestors
4. Donald M. Scott, "The Religious Origins of Manifest Destiny," Divining America, Teacher Serve©, National Humanities Center, http://nationalhumanitiescenter.org/tserve/nineteen/nkeyinfo/mandestiny.htm.
5. Anthony D. Barnosky and five hundred others, "Scientific Consensus on Maintaining Humanity's Life Support Systems in the 21st Century: Information for Policy Makers," 2013, http://mahb.stanford.edu/consensus-statement-from-global-scientists.

NOTES TO PART III

Moving with the Rhythm of Life

1. Dépaysement. In: *Dictionnaire de français Larousse* [online] Retrieved April 9, 2019, from https://www.larousse.fr/dictionnaires/francais/.
2. Stephen J. Pyne, *The Great Plains: A Fire Survey* (Tucson: University of Arizona Press, 2017).
3. Frances Weller, conversation with author, March 1, 2019.
4. Elissa Melaragno, "Trauma in the Body: Interview with Dr. Bessel Van Der Kolk," *Anchor*, November 18, 2015.

NOTES TO PART IV

Restoring Indigenous Mindfulness within the Commons of Human Consciousness

1. Paul Shepard, *Coming Home to the Pleistocene* (Washington, DC: Island Press, 1998), p. 38.
2. Alvin Josephy, ed., *America in 1492* (New York: Alfred A. Knopf, 1992), p. 7.
3. Ibid., p. 251.

NOTES TO PART V

The City Bleeds Out

1. Lao Tzu, *Tao Te Ching*, trans. Jonathan Star (New York: Jeremy P. Tarcher/Penguin, 2001), verse 78.
2. Lao Tzu, *Tao Te Ching*, verse 32.
3. Lao Tzu, *Tao Te Ching*, verse 4.

4. Ravens, for example, have captured a great deal of attention from cultures around the world. See Eric Mortensen, "Raven Augury from Tibet to Alaska: Dialects, Divine Agency, and the Bird's-Eye View," in *A Communion of Subjects: Animals, Science, and Ethics*, ed. Paul Waldau and Kimberley Patton (New York: Columbia University Press, 2006), 422–36. To delve deeper into avian esoterica, see Adele Nozedar, *The Secret Language of Birds: A Treasury of Myths, Folklore & Inspirational Stories* (London: HarperElement, 2006).

5. J. Freedman, "Iban Augury," *Bijdragen tot de Taal-, Land- en Volkenkunde* 117, no. 1 (1961): 141–67. According to Freedman, while other birds carry cultural significance, seven are most important to the Iban: rufous piculet, banded kingfisher, scarlet-rumped trogon, Diard's trogon, crested jay, maroon woodpecker, and white-rumped shama. Each has personality qualities unique to the bird and an elaborate symbology of associations related to color, habitual behavior, and types of vocalizations. "When any of these augural birds is seen or heard (so runs the theory of Iban augury) it can be assumed that the gods have something to communicate to man, for these birds, say the augurs, never reveal themselves without cause, they always have something to tell us" (147).

6. Freedman, 158.

7. Lao Tzu, *Tao Te Ching*, verse 8.

I Want the Earth to Know Me as a Friend

1. Enrique Salmón, "Kincentric Ecology: Indigenous Perceptions of the Human-Nature Relationship," *Ecological Applications* 10, no. 5 (2000): 1327–32.

2. Ibid.

Humus

1. Donna Haraway, "Anthropocene, Capitalocene, Plantationocene, Chthulucene: Making Kin," *Environmental Humanities* 6 (2015): 159–65, on p. 161.

2. Ibid, 160.

3. See Ashlee Cunsolo and Karen E. Landman, eds., *Mourning Nature: Hope at the Heart of Ecological Loss and Grief* (Montreal: McGill-Queen's University Press, 2017).

4. Louise Squire, "The Thoughts in our Head: A World," *Alluvium* 3, no. 1 (September 2014): n. pag. See also *The Environmental Crisis Novel: Ecological Death-Facing in Contemporary British and North American Fiction* (London: Routledge, 2019).

5. Lee Edelman, *No Future: Queer Theory and the Death Drive* (Durham, NC: Duke University Press, 2004).

6. J. Jack Halberstam, *In a Queer Time and Place: Transgender Bodies, Subcultural Lives* (New York: NYU Press, 2005).

7. I might also have chosen vermicompost. As Adam Phillips describes in his beautiful book *Darwin's Worms* (New York: Basic Books, 2001), after so much conflict in his life about the decentering of the human inherent in his evolutionary theory, Darwin found great hope for the world in the transformative work of worms. See also Sebastian Abrahamsson and Filippo Bertoni, "Compost Politics: Experimenting with Togetherness in Vermicomposting," *Environmental Humanities* 14 (2014): 125–48.

8. Catriona Sandilands, "Fear of a Queer Plant," *Gay and Lesbian Quarterly* 23, no. 3 (2017): 420.

NOTES TO PART VI

Seeds

1. Melissa K. Nelson and Nicola Wagenberg, "Linking Ancestral Seeds and Waters to the Indigenous Places We Inhabit," in *Ecological and Social Healing: Multicultural Women's Voices*, edited by Jeanine M. Canty, 100–120 (New York: Routledge, 2017).

2. See Paul C. Mangelsdorf, "The Role of Pod Corn in the Origin and Evolution of Maize," *Annals of the Missouri Botanical Garden* 35, no. 4 (1948): 377–406. doi:10.2307/2394701. https://www.jstor.org/stable/2394701?seq=1#page_scan_tab_contents

Landing

1. *Mosquita y Mari* (2012) is a coming of age film by Aurora Guerrero that follows two queer Chicanas who live in Huntington Park.

2. La Montaña refers to the mountain of concrete rubble in Huntington Park that appeared after the Northridge Earthquake in 1994. After community organizing, the mountain was removed from the site in 2001. Articles such as the following can be found on the topic and the work of community leaders. See http://articles.latimes.com/2001/may/01/local/me-57932

3. Robin Wall Kimmerer, *Braiding Sweetgrass: Indigenous Wisdom, Scientific Knowledge, and the Teachings of Plants* (Minneapolis: Milkweed Editions, 2013).

4. While we see the continuous attacks on the Standing Rock Sioux Tribe, I also acknowledge the resistance of our relatives in unceded Canadian territories that are currently asserting their existence against the Kinder Morgan Pipeline project.

5. Bears Ears is recognized a United States National Monument and stands on the lands of the Navajo Nation, Hopi, Ute Mountain Ute, Ute Indian Tribe of the Uintah and Ouray Reservation, and the Pueblo of Zuni. On December 4, 2017, President Donald Trump announced the monument would be reduced from its original 1,351,849 acres to 201,876.

6. Winnemem Wintu are the original peoples of the Mt. Shasta and McCloud River watershed. The Shasta dam currently occupies their land and was established in a violent flooding and loss of their salmon. They are currently protecting their lands from future expansions and plan to retrieve some of their lost Salmon from New Zealand. For more information, see http://www.winnememwintu.us

7. Sandy Grande, *Red Pedagogy: Native American Social and Political Thought* (Lanham: Rowman & Littlefield, 2015).

8. Mishuana Goeman, "Land as Life: Unsettling the Logics of Containment," *Native Studies Keywords* (Tucson: University of Arizona Press, 2015).

9. "Original Instructions refer to the many diverse teachings, lessons, and ethics expressed in the origin stories and oral traditions of Indigenous Peoples." Melissa K. Nelson, *Original Instructions: Indigenous Teachings for a Sustainable Future* (Rochester, VT: Bear and Co., 2008).

10. See http://www.thefeministwire.com/2012/10/the-shape-of-my-impact/.

11. Deborah A. Miranda, *Bad Indians: A Tribal Memoir* (Berkeley: Heyday, 2013).

12. Translates to chilies, although we mostly mean peppers like cayenne and other small peppers when referring to them.

13. Enrique Salmón, *Eating the Landscape: American Indian Stories of Food, Identity, and Resilience* (Tucson: University of Arizona Press, 2012).

14. This was a campaign spearheaded by CBE where residents of East Oakland took action against a crematorium in their neighborhood. Although there was an understanding it was defeated, months into 2017, the crematorium began operations. See https://www.eastbayexpress.com/oakland/the-return-of-the-crematorium/Content?oid=10841726.

15. I was introduced to the stories of the Shellmounds and the incredible action being taken to protect them from Corrina Gould of Indian People Organizing for Change (IPOC). This group worked with an all-Indigenous-women-led land trust. For information on the Shellmounds see http://ipocshellmoundwalk.homestead.com/shellmound.html. For information on the land trust see http://sogoreate-landtrust.com.

16. I take from Jose Esteban Muñoz's (1999) concept of queer worldmaking where he asserts, "It is my contention that the *doing* that matters most and the performance that seems most crucial are nothing short of the actual making of worlds."

17. While I ground my analysis in haunting on the work of Avery Gordon (1997), this definition is particularly inspired by the work of Eve Tuck and C. Ree in "A Glossary of Haunting," in *Handbook of Autoethnography*, ed. Stacy Linn Holman Jones, Tony E. Adams, and Carolyn Ellis (London: Routledge, Taylor & Francis Group, 2013).

Regenerative

1. This is a term I first heard from Sakej Henderson to refer to the Native youth who went to boarding or residential school in US and Canada and had to "split their heads" between being Indian and being assimilated into white society. I've written about it in *Original Instructions: Indigenous Teachings for a Sustainable Future* (Rochester, VT: Bear and Co., 2008).

2. See Eduardo Duran's *Healing the Soul Wound: Counseling with American Indians and Other Native Peoples* (New York: Teachers College Press, 2006) and Eduardo Duran and Bonnie Duran, *Native American Postcolonial Psychology* (Albany: State University of New York Press, 1995).

3. From John Borrows, *Drawing Out Law: A Spirit's Guide* (Toronto: University of Toronto Press, 2010), p. 105.

4. Eighth Fire draws from an Anishinaabe prophecy that declares now is the time for Aboriginal/Indigenous peoples and the settler community to come together and build the "Eighth Fire" of justice and harmony. (http://www.cbc.ca/8thfire/index.html).

5. I first heard this expression from attorney Claire Cummings in 1993 where she and the Cultural Conservancy held a meeting under that title. The phrase was later used by Darrell Posey in a major article, "Commodification of the Sacred through Intellectual Property Rights" (2002).

6. A line from Robinson Jeffers from the preface to *Be Angry at the Sun (and Other Poems)* (New York, 1941).

7. 2012 Ted Talk, YouTube.

Interview: Ilarion Merculieff

1. Larry Merculieff, *Wisdom Keeper*: One Man's Journey to Honor the Untold History of the Unangan People (Berkeley: North Atlantic Books, 2016), p. 28.

2. To learn more about similar practices, please refer to the following film: *Atanarjuat: The Fast Runner*, filmed by Zacharias Kunuk and produced by Isuma Igloolik Productions, 2001.
3. See www.wisdomweavers.world.
4. *Wisdom Keeper*, p. 161.

*

Special thanks goes to our manuscript editor, Mary Corrado, our indexer, Jim Fuhr, and our final proof assistant, Samuel Archibald-Gutshall.

ABOUT THE CONTRIBUTORS

Aaron A. Abeyta is a Colorado native and professor of English and the mayor of Antonito, Colorado, his hometown. He is the author of four collections of poetry and one novel. For his book *colcha*, Abeyta received an American Book Award and the Colorado Book Award. In addition, his novel, *Rise, Do Not Be Afraid*, was a finalist for the 2007 Colorado Book Award and El Premio Aztlan.

Leah Bayens is dean of the Wendell Berry Farming Program of Sterling College, based in Henry County, Kentucky. She works with colleagues and community partners to translate Wendell Berry's visions for thriving farming communities into hands-on, interdisciplinary, new agrarian education. In 2011, Leah earned a doctorate in English from the University of Kentucky, where she studied nineteenth-century American literature, ecocriticism, and cultural studies theory. Her research in agrarian literature, history, and culture has been published in *Working for Social Justice: Inside and Outside the Classroom*, *Appalachian Heritage*, *The Whole Horse Project*, and *The Notebook: A Journal for Rural Women and Girls*. She lives on family land near Danville, Kentucky, with her partner Bruce Bryant, a cabinetmaker and part-time teacher, and their son, Burley.

Wendell Berry is a Kentucky writer and farmer who has authored more than fifty works of fiction, nonfiction, and poetry. Berry was born in Henry County, Kentucky, in 1934, to Virginia Erdman and John Marshall Berry, Sr. His father, a lawyer and farmer, was a principal author of the Burley Tobacco Producer's Program, which protected farmers in the marketplace. Like his predecessors, Berry has devoted his life to small farmers and land-conserving communities, as evidenced in *The Unsettling of America: Culture and Agriculture*, a landmark book that has sparked over four decades of conversation about the state of agriculture. Berry and his wife, Tanya, have lived and worked on their 120-acre farm in Port Royal, Kentucky, since 1964. They have two children, Mary and Den. Mary Berry is the executive director of the Berry Center, and Den Berry is a Henry County farmer who serves on the Berry Center's Board of Directors.

About the Contributors

Kaylena Bray (Haudenosaunee/Seneca) is Turtle Clan from the Seneca Nation of Indians. She has grown up eating traditional white corn, which has given fuel to a career focused on strengthening Indigenous knowledge of traditional agriculture, Native foodways, and environmental health. Her work throughout the Americas has served to educate and strengthen vital links between Indigenous food systems, local economies, and climate change adaptation. She holds degrees from Brown University and the University of Oxford, and currently supports small scale funding for traditional farming and local economic development initiatives throughout Turtle Island.

Brian Calvert holds an MFA in poetry from Western Colorado University and was awarded a Ted Scripps Fellowship at UC Boulder's Center for Environmental Journalism.

Taiyon Coleman is a Cave Canem and VONA fellow, and her writing has appeared in *Bum Rush the Page: A Def Poetry Jam*; *Riding Shotgun: Women Writing about Their Mothers*; *The Ringing Ear: Black Poets Lean South* ; *Blues Vision*; *How Dare We!, Write: A Multicultural Creative Writing Discourse*; *Time for the Truth: Race in Minnesota*; *Places Journal*; and *Minnesota*. Taiyon has essays in the forthcoming *Shadowlands: An Illustrated Reader in Racialized Violence in America, Selected Writings on the Art of Ken Gonzales-Day* (Minnesota Museum of American Art) and *What God Is Honored Here? Writings on Miscarriage and Infant Loss by and for Native Women and Women of Color* (University of Minnesota Press). Taiyon is a 2017 recipient of a McKnight Foundation Artist Fellowship in Creative Prose, and she is one of twelve Minnesota emerging Children's Writers of Color selected as a recipient of the 2018–2019 Mirrors and Windows Fellowship funded by the Loft Literary Center and the Jerome Foundation. Taiyon is assistant professor of English Literature at St. Catherine University in St. Paul, Minnesota.

Katherine Kassouf Cummings serves as managing editor at the Center for Humans and Nature and leads Questions for a Resilient Future.

Camille T. Dungy is author most recently of the poetry collection *Trophic Cascade* and a collection of personal essays called *Guidebook to Relative Strangers*. She is also an editor of anthologies, including *Black Nature: Four Centuries of African American Nature Poetry*, and professor of English at Colorado State University.

Peter Forbes is the author and photographer of five books about people living in place in North America. He does that himself in his beloved northern New England, where he and his family steward Knoll Farm, a regenerative sheep and berry farm in the Mad River Valley of Vermont. Peter also builds community by being a cross-cultural facilitator on the journey to strengthen rural communities through courageous convening across differences of race, class, gender and ideology. peterforbes.org.

Kristi Leora Gansworth is Anishinaabe-kwe, a poet who works in the field of human geography. She is a citizen of Kitigan Zibi Anishinaabeg, and her work grows from experience and lived knowledge.

Shannon Gibney is a writer, educator, and activist. She is the author of the novels *Dream Country* and *See No Color*, which both won Minnesota Book Awards. Most recently, she edited the anthology *What God is Honored Here? Writings on Miscarriage and Infant Loss by and for Native Women and Women of Color*, with writer Kao Kalia Yang. Gibney is faculty in English at Minneapolis College.

About the Contributors

Oscar Guttierez is a PhD student in the Department of Ethnic Studies at the University of California, San Diego. His work is at the intersections of critical Latinx and Indigenous studies and geography. Currently, he writes about issues of environmental justice and memory in Southeast Los Angeles, where he is from. Oscar works closely with Communities for a Better Environment, an environmental health and justice organization where he is currently a board member.

John Hausdoerffer is an environmental and social justice thinker living in Gunnison, Colorado. He is author of *Catlin's Lament: Indians, Manifest Destiny, and the Ethics of Nature*; the editor of Aaron Abeyta's *Letters from the Headwaters*; and coeditor and coauthor of *Wildness: Relations of People and Place*. He is currently coediting a new book entitled *Kinship: Belonging in a World of Relations*. John serves as the dean of the School of Environment & Sustainability at Western Colorado University. John is the cofounder of numerous organizations, including the Mountain Resilience Coalition, the Resilience Studies Consortium, the Coldharbour Institute, and Western's Master in Environmental Management Program. In 2018, Peace Museum Colorado named him a "Peace Hero." jhausdoerffer.com.

Brooke Parry Hecht Seeking the space to explore life's big questions, Brooke joined the Center for Humans and Nature in 2005 as a research associate. She has been the president of the Center since 2008. Whether through the Center's Questions for a Resilient Future program or other Center initiatives, her work explores what it means to be human and what our responsibilities are to each other and the whole community of life.

Elizabeth Carothers Herron writes poetry and articles on art and ecology, the role of art in society, and the importance of natural systems and biodiversity in the physical and spiritual well-being of individuals, communities, and the planet. Much of her poetry reflects these concerns. Her collection of poems, *Insistent Grace*, is forthcoming this year from Fernwood Press. elizabethherron.net.

Linda Hogan (Chickasaw) is former writer-in-residence of the Chickasaw Nation and professor emerita from the University of Colorado and is an internationally recognized public speaker and writer of poetry, fiction, and essays. Her books include *Dark Sweet—New and Selected Poems* (2014) and *Indios* (Wings Press 2012). Her fiction work includes the novel *Mean Spirit*, a winner of the Oklahoma Book Award and the Mountains and Plains Book Award, and a finalist for the Pulitzer Prize.

Wes Jackson is a geneticist, author, and visionary leader of the sustainability movement. He was first to chair the Environmental Studies Department at California State University-Sacramento, one of the first such programs in the US. He left that position to return to his home state of Kansas, where he and his then-wife, Dana, cofounded The Land Institute. There he originated the concept of transforming agriculture to mimic the prairie, with development of perennial grains grown in polyculture. Wes is a recipient of the Pew Conservation Scholars award (1990), the MacArthur Fellowship (1992), and the Right Livelihood Award (Stockholm), known as the "Alternative Nobel Prize" (2000). He continues to work on behalf of TLI's Ecosphere Studies program, which explores the broader social, political, and economic changes necessary to allow both agriculture and culture to flourish.

Princess Daazhraii Johnson is interested in systemic change for a more just and equitable world. Princess is an actor/ filmmaker/ writer and is a Sundance Directing Fellow and an Emerging Voices/PEN Center alumni. Princess received a BA in international relations from the George Washington University and a MEd at the University of Alaska Anchorage with a focus on environmental and science education. In 2015 she was appointed by President Obama to serve on the Board of Trustees for the Institute of American Indian Arts. She is also honored to serve on the Boards of NDN Collective and Native Movement and is a founding member of the Fairbanks Climate Action Coalition. She lives in Fairbanks, Alaska, with her husband, James, and her three boys and is currently creative producer and writer of the PBS Kids show called *Molly of Denali*.

Lyla June Johnston is a poet, musician, human ecologist, public speaker, and community organizer of Diné (Navajo), Tsétsêhéstâhese (Cheyenne), and European lineages. Her dynamic, multi-genre performance style has invigorated and inspired audiences across the globe towards personal, collective, and ecological healing. Her messages focus on Indigenous rights, supporting youth, intercultural healing, historical trauma, and traditional land stewardship practices. Her undergraduate studies in human ecology at Stanford University, mixed with surviving trauma of many kinds and learning traditional Native medicines, inform her potent messages conveyed through the medium of powerful yet prayerful hip-hop, poetry, acoustic song, and speech. She is a cofounder of the Taos Peace and Reconciliation Council, which works to heal intergenerational trauma and ethnic division in northern New Mexico. She was a walker within the Journey for Our Existence Movement, a 1400-mile prayer journey through Diné homelands for environmental and social justice. She is the lead organizer of the Black Hills Unity Concert. She is the also the founder of Regeneration Festival, an annual celebration of children that occurs in 13 countries around the world every September. In 2012, she graduated with honors from Stanford University with a degree in environmental anthropology. She recently graduated from the University of New Mexico with distinction with a concentration in American Indian Education. She is pursuing her PhD at the University of Alaska Fairbanks.

Frances H. Kakugawa Award-winning, internationally published author/ poet of more than a dozen books for adults and children, Frances also writes a Dear Frances advice column for caregivers in the *Hawai'i Herald*. She conducts lectures/ workshops on respite care, poetry, and education for children and adults. Frances, whose childhood village was destroyed by lava flows in Hawaii, currently resides in Sacramento, California.

Robin Wall Kimmerer is a mother, scientist, writer, and an enrolled member of the Citizen Band Potawatomi. Her writings include numerous scientific articles and the books *Gathering Moss*, which was awarded the John Burroughs Medal for nature writing in 2005, and *Braiding Sweetgrass: Indigenous Wisdom, Scientific Knowledge and the Teachings of Plants*, which was awarded the Sigurd Olson Nature Writing Award. She is a Distinguished Teaching Professor of Environmental Biology at the SUNY College of Environmental Science and Forestry in Syracuse, New York, and the founding director of the Center for Native Peoples and the Environment, whose mission is to create programs that draw on the wisdom of both Indigenous and scientific knowledge for our shared concerns for Mother Earth. She lives on an old farm in upstate New York, tending gardens both cultivated and wild.

About the Contributors 267

Winona LaDuke lives and works on the White Earth reservation in northern Minnesota, and is a two-time vice presidential candidate with Ralph Nader for the Green Party. As program director of Honor the Earth, she works nationally and internationally on the issues of climate change, renewable energy, and food systems. In her own community, she is the founder of the White Earth Land Recovery Project. She has authored *Recovering the Sacred*, *All Our Relations*, and *Last Standing Woman*. In 2007, LaDuke was inducted into the National Women's Hall of Fame. Her book *Chronicles* came out in 2017 and her new book, *To Be a Water Protector*, came out in autumn 2020 with Fernwood Press.

Estella Leopold is a University of Washington professor emerita of botany, forest resources, and quaternary research. She has been teaching and conducting research for more than sixty years. The author of more than one hundred scientific publications in the fields of paleobotany, forest history, restoration ecology, and environmental quality, Estella pioneered the use of fossilized pollen and spores to understand how plants and ecosystems respond over eons to climate change and other phenomena. She is a Fellow of the American Association for the Advancement of Science, the Geological Society of America, the American Academy of Arts and Sciences, and the American Philosophical Society, and was elected to the US National Academy of Sciences. With her four siblings, she established the Aldo Leopold Foundation in Baraboo, Wisconsin.

Jack Loeffler is an aural historian, author, radio producer, and lecturer who has engaged in environmental activism for over fifty years. He is the author of nine books, and producer of over four hundred documentary radio programs broadcast over Community Public Radio stations throughout the Southwest and beyond. He forwards systems thinking in an effort to clarify an environmental ethic in both his writing and radio production. In 2019, his memoir entitled *Headed into the Wind* was released by the University of New Mexico Press.

Lindsey Lunsford is a scholar activist and agriculture advocate, inspired by her alma mater, Tuskegee University, who is setting the tone for the upcoming generation of urban agriculturalists. Currently working as Sustainable Food System Resource Specialist through Tuskegee University's Carver Integrative Sustainability Center, Dr. Lunsford works in food justice and community sovereignty efforts. Lunsford recently completed the Tuskegee University Integrative Public Policy and Development Doctoral Program, where she studied policy advocacy for strengthening grassroots efforts from state to international levels. Lunsford's leadership focuses on educating urban youth on self-sufficiency and healthy lifestyles while encouraging the larger community to pursue a sustainable local economy.

Jamaal May is a writer, teacher, mixed-media artist, and member of the Organic Machine Nation. His creative projects explore the tension between polarities to render an argument for the interconnectivity of people, worlds, and ideas. He has exhibited at the Smithsonian and published two award-winning poetry collections: *Hum* and *The Big Book of Exit Strategies* (Alice James Books, 2013, 2016). Honors include the Spirit of Detroit Award, the Benjamin Danks Award from the American Academy of Arts and Letters, an NAACP Image Award nomination, and a Lannan Foundation grant. Jamaal has worked as a hotel "a.v. guy," a freelance audio engineer/producer, and an Inside Out Writer in Residence in Detroit Public Schools. He is currently an assistant professor at Wayne State University.

From Hamtramck and Detroit, Michigan, he directs the Organic Machine Books, the Blair Prize, and the Qualia Bridge Project.

Christopher (Toby) McLeod has been project director of Earth Island Institute's Sacred Land Film Project since 1984. In 2013, he completed the four-part series Standing on Sacred Ground, which premiered on public television in May 2015. Previously, Toby produced and directed *In the Light of Reverence* (2001) and made three other award-winning, documentary films that were broadcast on national television: *The Four Corners: A National Sacrifice Area?* (1983), *Downwind/Downstream* (1988), and *NOVA: Poison in the Rockies* (1990). In 1997, he completed *A Thousand Years of Ceremony*, a 40-minute profile of Winnemem Wintu healer Florence Jones and her efforts to protect Mount Shasta as a sacred site—a film made specifically as an archival film for use by the Winnemem. Toby has been working with Indigenous communities as a filmmaker, journalist, and photographer for forty years.

Curt Meine is a conservation biologist and writer based in Sauk County, Wisconsin. He serves as senior fellow with the Aldo Leopold Foundation and the Center for Humans and Nature; as research associate with the International Crane Foundation; and as adjunct associate professor at the University of Wisconsin-Madison. He served as on-screen guide in the Emmy Award–winning documentary film *Green Fire: Aldo Leopold and a Land Ethic for Our Time* (2011). Meine has authored and edited several books, including the award-winning biography *Aldo Leopold: His Life and Work* (2010) and *The Driftless Reader* (2017). In his home landscape, he is a founding member of the Sauk Prairie Conservation Alliance.

Ilarion Merculieff Close to Ilarion (Larry) Merculieff's heart are issues related to cultural and community wellness, traditional ways of living, Elder wisdom, and the environment. Having had a traditional upbringing, Merculieff has been, and continues to be, a strong voice advocating the meaningful application of traditional knowledge and wisdom obtained from Elders in Alaska and throughout the world when dealing with modern-day challenges.

Kathleen Dean Moore is an environmental philosopher and climate advocate—the author or coeditor of a dozen books in celebration and defense of the reeling world. The most recent are *Moral Ground*, *Great Tide Rising*, and *Piano Tide*, winner of the Willa Cather Award for Contemporary Fiction. Until recently Distinguished Professor of Philosophy at Oregon State University, Moore now writes and speaks full-time about the moral urgency of climate action. With pianist Rachelle McCabe, she performs "A Call to Life," a music and spoken-word concert about the global extinction crisis. She is currently working on a book about the human-rights impacts of fracking and climate change. riverwalking.com, musicandclimateaction.com.

Melissa K. Nelson (Anishinaabe/Métis [Turtle Mountain Chippewa]) is an ecologist, writer, and Indigenous scholar-activist. She is a professor of American Indian Studies at San Francisco State University and president of the Cultural Conservancy, a Native-led Indigenous rights organization, which she has directed since 1993. Melissa is the editor of and contributor to two books, *Original Instructions—Indigenous Teachings for a Sustainable Future* (2008) and *Traditional Ecological Knowledge: Learning from Indigenous Practices for Environmental Sustainability* (2018).

Sean Prentiss, his wife, and his daughter find home on a small lake in northern Vermont. He is the award-winning author of *Finding Abbey: The Search for Edward Abbey and His Hidden Desert Grave* and *Crosscut: Poems*. He is an associate professor at Norwich University and a core faculty at Vermont College of Fine Arts.

Enrique Salmón is a Rarámuri (Tarahumara) Indian. He has a PhD in anthropology from Arizona State University. He is head of the American Indian Studies program at Cal State University East Bay. Enrique has published several articles and chapters on Indigenous ethnoecology, agriculture and ancestral food ways, nutrition, sustainability education, and traditional ecological knowledge. Dr. Salmón is author of the books *Eating the Landscape: American Indian Stories of Food, Identity, and Resilience* and *Iwígara: American Indian Ethnobotanical Traditions and Science*.

Catriona Sandilands is a professor of environmental humanities in the Faculty of Environmental Studies at York University and a fellow of the Pierre Elliott Trudeau Foundation. She is known for her extensive writing on queer and feminist ecologies, including the coedited anthology (with Bruce Erickson) *Queer Ecologies: Sex, Nature, Politics, Desire* (Indiana 2010), and for her current research on plants and botanical relationships. She has also edited a volume of creative writing on climate change, *Rising Tides*, published by Caitlin Press in 2019.

Vandana Shiva trained as a physicist at the University of Punjab and completed her PhD at the University of Western Ontario, Canada. She later shifted to interdisciplinary research in science, technology, and environmental policy. Dr. Shiva has contributed in fundamental ways to changing the practice and paradigms of agriculture and food. Her books *The Violence of the Green Revolution* and *Monocultures of the Mind* pose essential challenges to the dominant paradigm of nonsustainable, industrial agriculture. Through her books *Biopiracy*, *Stolen Harvest*, and *Water Wars*, Dr. Shiva has made visible the social, economic, and ecological costs of corporate-led globalization. In November 2010, *Forbes* identified Dr. Shiva as one of the "Seven Most Powerful Women on the Globe." Dr. Shiva advises governments worldwide and is currently working with the Government of Bhutan to make Bhutan 100 percent organic.

Caleen Sisk is the Chief and Spiritual Leader of the Winnemem Wintu Tribe, who practice their traditional culture and ceremonies in their territory along the McCloud River watershed in Northern California. Since assuming leadership responsibilities in 2000, Caleen has focused on maintaining the cultural and religious traditions of the tribe, and has led the revitalization of the Winnemem's H'up Chonas (War Dance) and BaLas Chonas (Puberty Ceremony), which had not been practiced for decades. She advocates for California salmon restoration, healthy, undammed watersheds, and the human right to water. She has received international honors as a tireless sacred site protector, and currently leads the tribe's resistance against the US Bureau of Reclamation's proposal to raise Shasta Dam by 18.5-feet, which would inundate or damage more than 40 sacred sites. Strongly rooted in her spirituality and her family, Caleen cares deeply for her Winnemem people and for oppressed people around the world.

Chellie Spiller is a professor at the University of Waikato Management School. She is a passionate and committed advocate for Māori and Indigenous leadership, management, governance, and business development. She is a coauthor of *Wayfinding Leadership* with Hoturoa Barclay-Kerr and John Panoho. See her Wayfinding Leadership Tedx www.youtube.com/watch?reload=9&v=d1-gmU04jhs. Chellie has also coedited a book on *Authentic Leadership* with Professor Donna Ladkin and a book on *Indigenous Spiritualities at Work* with Dr. Rachel Wolfgramm.

Aubrey Streit Krug is a writer and teacher who studies stories of relationships between humans and plants. She directs the Ecosphere Studies program at The Land Institute in Salina, Kansas, where she lives with her husband and son. She grew up in a small town in Kansas, where her parents farm wheat and raise cattle, and she loves limestone soils and rocky prairie hillsides. Streit Krug holds a PhD in English and Great Plains Studies from the University of Nebraska-Lincoln and is a coauthor of the collaborative textbook *The Omaha Language and the Omaha Way*.

Manea Sweeney is an environmental director, urban planner, and writer from Wellington, New Zealand. Over the past ten years she has worked in disaster resilience and recovery, public works, and community engagement projects, with a focus on the importance of Indigenous knowledge and voices to respond to tomorrow's world.

Gavin Van Horn works at the Center for Humans and Nature. He is the coeditor of *City Creatures: Animal Encounters in the Chicago Wilderness* and *Wildness: Relations of People and Place*. Gavin leads interdisciplinary projects for the Center that explore place-based values and human relationships to the more-than-human world. He writes for, edits, and curates the City Creatures Blog and his words have appeared in *Orion, Emergence Magazine, Belt Magazine, Red Savina Review*, and *Zoomorphic*, among others. His most recent book is *The Way of Coyote: Shared Journeys in the Urban Wilds*.

Nicola Wagenberg, PhD, is a clinical and cultural psychologist and educator. She has worked for over 20 years with diverse individuals, communities, and organizations on personal and cultural transformation. Since 2005, Nicola has been working with the Cultural Conservancy directing media projects, and developing and implementing arts and cultural health programs. Nicola is also a psychotherapist in private practice where she sees individuals, couples, and groups in Berkeley, California.

Julianne Lutz Warren is a freelance storyteller and community co-organizer with an MA in linguistics and a PhD in ecology. As she is of European lineages and settler-colonial birth-culture, her creative pieces, including sound arts, are about re/learning what it is to be in good relations. Works appear in a variety of venues, including *Newfound, Minding Nature, Zoomorphic*, the Poetry Lab of the Merwin Conservancy, Lost and Found Theatrum Anatomicum, and The Deutsches Museum. As a faculty member, Julianne collaborated with students growing NYU Divest: Go Fossil Free! While living far north, she served as a council member of Fairbanks Climate Action Coalition and cofacilitator of #KeepItIntheGround! working group. She is a named Scholar and Fellow at the Center for Humans and Nature and an Ecosphere Studies collaborator and visiting scholar at The Land Institute. She currently resides in ancestral Tewa homelands of Northern New Mexico.

Rowen White is a Seedkeeper and farmer from the Mohawk community of Akwesasne and a passionate activist for indigenous seed and food sovereignty. She weaves stories of seeds, food, culture, and sacred Earth stewardship on her blog, Seed Songs, and cultivates a legacy of seeds and cultural memory with the Indigenous Seedkeepers Network.

Brooke Williams' life has been one of adventure and wilderness exploration. His conservation career spans forty years. His most recent book, *Open Midnight*, documents his exploration of places where the outer and inner wilderness meet. He's now writing about dragonflies and hermits. He believes that the length of the past equals the length of the future.

Crystal Williams A poet and essayist, whose most recent book is *Detroit as Barn*, Crystal Williams is associate provost for diversity and inclusion and professor of English at Boston University.

Eryn Wise comes from the Jicarilla Apache Nation and Pueblo of Laguna. She is the communications and digital director at Seeding Sovereignty, and is a cofounding director of the Indigenous Impact Rapid Response Initiative. Ensuring a future for generations to come is a duty and responsibility to her, and she honors the traditional teachings of her predecessors by continuing to build intersectional communities in the spirit of kinship and Indigenous resurgence. She aims to do all her work in homage to her ancestors, who continue to empower and inspire her journey. She is a human being, just like you.

Rachel Wolfgramm (Whakatōhea, Ngai Takoto, Te Aupouri, Tonga) is a principal investigator for Nga Pae o te Maramatanga and is currently leading a project along with a team of senior Maori academics and doctoral students investigating leadership in economies of wellbeing. She is a senior lecturer at the University of Auckland Business School and is an active researcher, author, and consultant in sustainability, leadership, intercultural communications, and Māori development. Over the past 15 years, her research has been published in international journals and books and presented at numerous conferences across Europe, USA, and Asia Pacific.

INDEX

Page numbers in **_bold italics_** indicate illustrations.

Abenaki Tribe/Nation, 174
Abeyta, Aaron A., 11, 16–19, 263, 265
acid rain, 128
Adams, John Quincy, 62
African Americans and Black people: and chronic joyfulness, 111, 113; considered less than human, 57; as Diaspora, 113–14, 116; DNA as sole legacy of, 58; Earth respected by, 59; as farmers, 58, 64–65; farms stolen from, 58; and gun ownership, 65; infertilization of, 60; labor unions run by, 56–57; targeted for underdevelopment, 58; and the vocalization of appreciation, 112; women ancestors of, 110. *See also* Coleman, Taiyon; Dungy, Camille T.; Lunsford, Lindsey; Williams, Crystal
agriculture: and ancestry, 166; Indigenous knowledge of, 264, 269; and The Land Institute, 265; Lindsey Lunsford's advocacy for, 267; Pleistocene development of, 132, 136; as scourge, 65; Shiva's activism regarding, 269; and skin color, 64–65; *The Unsettling of America*, 263; and white supremacy, 65. *See also* farms and farming

air pollution, 43, 84, 191, 214
Alaska: going barefoot in, 240–41; in *Drums of Winter*, 108; Gwich'in First Nations people of, 108; "The Mystery of the Alaskan Mummies," 246; Russian oppression in, 241, 243–44, 246; slavery in, 243–44; Unangan islands and people of, 240–50. *See also* Merculieff, Ilarion
Aldo Leopold Foundation and Center for Humans and Nature (Baraboo, Wisconsin), 139, 267–68
Aleutian Islands (Alaska), 240, 242–43, 246. *See also* Unangan (Aleutian) Islands and people
Alexie, Sherman, 218
"Alive in This Century" (Gansworth poem), 121–23
American Indians. *See* Indians; Indigenous people; Native Americans
American Petroleum Institute, 71
American Tobacco Company, 34
Amistad (film), 62–63
amnesia, 91, 143, 250
"Ancestor of Fire" (Abeyta essay), 16–19

ancestors: all beings as, 189–90; in Anishinaabe, to string/tie together, 220; and corn, 200, 210; criteria for, 166; DNA (Descendants-N-Ancestors), 3, 220; as ecological change-makers, 145; embedded in their environments, 4; and the future, 216, 218; as heirs, 71; honoring the right ancestors, 14; "To Hope of Becoming Ancestors," 100–109; Jewish, adapting to new cultures, 142; Māori, encouraging cultural exploration/inquiry, 222–23, 238; multiplicities of, 220; of the Northern Plains, 219; overview, 1–5; personified in descendants, 206; in present time, 205; seen in many places, 218; soil built/nurtured by, 177–81, 182; stars as, 149–50; as tree-planters, 172; ubiquity of, 205; as warriors, 142; "What Kind of Ancestor Do You Want to Be? A Plantcestor!," 202–3; "What Kind of Ancestor Will I Be?," 201–2; without children, 126, 222. *See also* descendants; other-than-human beings; seeds

ancestral cycles, 31

ancestral foods, 169, 269

ancestral grounds/lands/soil: as ancestors of humans, 90; and dance, 90–94; descendants' relations with, 98; displacement from, 54, 80, 83, 90, 228; guardianship of, 208; left for future generations, 56; lives united with, 30; making and remaking of, 22–23; and manifest destiny, 66; residence on, 270; revealing human stewardship, 98–99; unceded, 108; unearthing of, 20

ancestral instruction/teaching, 188, 193, 217

ancestral legacies, 147, 208, 211, 229

ancestral responsibility, 115, 140–44, 179–80. *See also* "Embedded"

ancestral rights, 54

ancestral waters, 3, 27, 30, 80, 90

ancestral wisdom, 130

ancestry: and being alive in the present, 127; colonial disruption of, 4; continuing in healthy ecospheres, 153; control of, attempted, 105; and eco-cultural health, 208; ecological, 179; ethical, 140–41, 179; and farming, 64–65; and the future, 21, 70–72, 180, 185; and intergenerational accountability, 143; and the last universal common ancestor, 130; lost, 60, 105; in Māori genealogical narratives, 232–39; origins of, 61; as the past in the future, 21; in the present moment, 188; recorded, 138–41; and rice farming, 128–29; rituals honoring, 3; seed traditions of, 193; and soil, 4; space and time in, 126; vs transience, 144–45; unworthy practices in, 188–89; worthiness in, 104, 128, 183, 188–89, 220. *See also* family histories; genealogies; place-sense; seeds; soil

Anishinaabe(g) tribes/Nation, 4; *Akiing* (people/land unity) of, 142–43; ancestry (*aanikoobijigan*) of, 220; berries as gifts among, 261n4; Dahl as elder/Lodge Leader of, 1, 126; and the Eighth Fire, 221, 261n4; Gansworth as descendant of, 264; the good life (*mino bimaatisiiwin*) of, 143; Nelson as descendant of, 3; relationships with more-than-human beings, 23; as Seventh Fire people, 4, 187; way of life (*miikana*) of, 122; as wild rice culture, 190. *See also* Odawa Indians; Ojibwe People; Pot(t)awatomi Nations

Anthropocene epoch, 178, 180

anthropocentrism, 143, 163, 185–86, 191

anti-Semitism, 63

Aotearoa mountains (New Zealand), 107, 234–35

Apache Tribe. *See* Jicarilla Apache Nation

"Apple Tree, The" (Forbes essay), 172–76

Arel, Christian, 153

Arnhem (Australia), 166–67

Atanarjuat (film), 262n2

Australia, 166–67

Avalos, Benny, 201–2

Awiakta, Marilou, 225

Barclay-Kerr, Hoturoa, 270
Barlow, Cleve, 232
Barnes, Carl, 212
Bayens, Leah, 33–43, 263
Be Angry at the Sun (and Other Poems) (Jeffers), 261n6
Bears Ears National Monument (Utah), 215, 260n5
Beauty Way (Navajo traditional prayer), 223
Beck, Johannes Georg, 28–29
Beck lineage, 28
Benton-Benai, Edward, 195
Bering Sea, 241
Berry, Den, 36, 263
Berry, Mary, 43, 263
Berry, Tanya, 42, 263
Berry, Wendell, 33–43, 263
Berry Center, 43, 263
Bethe, Hans, 150
Big Bang, 145, 153
Big Heads (Winnemem Wintu spirit-teachers), 79, 83
Bihari Muslims, 187
biodiversity, 118, 146–47, 173, 184, 192, 265
bioregionalism, 135–36
birds as auguries, 163–64, 259nn4–5
Black Hills Unity Concert (South Dakota), 266
Black Mountain Circle (California), 255
Black people. *See* African Americans and Black people
Bluestem roots, 20–24
Bluffs, The (Canada), 177, 181
Bodwewadmi Nations. *See* Pot(t)awatomi Nations
Bohm, David, 221
Borneo, 163
Bray, Kaylena, 210–13, 264
Brotherhood of Sleeping Car Porters, 56–57
Brubaker, Ben, 42
Buddhism, 179, 187, 202
"Building Good Soil" (Kimmerer essay), 182–84

Buliyum Puyuk. *See* Mount Shasta (Buliyum Puyuk)
Burley Tobacco Growers Co-operative Association, 33
Burley Tobacco Producer's Program, 263
burning sage ceremony, 1
Burroughs, John, 102
Busse, James S., 21

California: drought in, 74; GOTW program in, 200, 205; Indians of, dispossessed, 83; Miwok homeland in, 1; native perennial plants of, 224; water pollution in, 82; Winnemem Wintu ceremonies in, 269
Calvert, Brian, 89, 264
Canada, 177, 191, 249–50, 261n1
Canadian Sioux, 260n4
capitalism, 69, 82, 180, 189, 192, 221–22
Capra, Fritjof, 131
carbon, 145–46, 149, 151, 160, 178
carbon dioxide, 186, 191
carnal (Latino concept of interconnectedness), 168
Carver, George Washington, 57
CBE (Communities for a Better Environment) (Los Angeles County), 214–15, 261n14, 265
Center for Humans and Nature (Chicago), 220, 255, 264, 268, 270
Center for Native Peoples and the Environment (Syracuse, New York), 266
Charles, Lala, 101–2, 106
Cheddar Man, 258n2
"Cheddar Man" (Williams essay), 61–67
Cherokee Nation, 212, 225
Cheyenne Tribe, 266
Chicago, 90–91, 159, 161, 164
Chicago River, 160
Chican@ (Chicano/Chicana ethnicity), 202
Chickasaw Delta, 13
Chickasaw Nation, 265
Chihuahua (Mexico), 141
China, 132
Chippewa Indians, 269

Choctaw Delta, 13
Choctaw Indians, 200, 206
Chödrön, Pema, 112
Christianity, 42, 52–53, 65–66, 149
Cinque (character in *Amistad*), 62–63
"City Bleeds Out (Reflections on Lake Michigan), The" (Van Horn), 159–65
clans, 54, 134, 191, 221, 264
Clark, Thomas D., 35
Clifton, Lucille, 112
climate change/disruption, 2, 92, 124–29, 182, 264, 267–68
cognition, 130–31, 136
Cold Mountain (Chinese poet), 26–30, 32
Cold Mountain (Vermont), 26–30
Coleman, Frank, 13
Coleman, Taiyon, 13–15, 264
colonialism and colonization: Americans as colonialists, 145–46, 151; and ancestry tracing, 218; attitudes of, unlearned, 103; Canadians as colonists, 177; the decolonized spirit vs, 221; decolonizers vs, 206–7, 216; descendants of colonizers, 270; healing from, 4, 87, 220–23, 225; in India, 191–92; of the Māori, 95, 98, 107; migration routes vs boundaries of, 54; and overconsumption, 1; religions of, 52; seed cultures disrupted by, 228; and white supremacy, 185; youth culture's rejection of, 223
Colorado, 123–25, 128–29, 134, 139
Colorado River, 139
Columbia River, 139
Columbus, Christopher, 66, 187, 189
commodification of the sacred, 221, 261n5
Communities for a Better Environment (CBE) (Los Angeles County), 214–15, 261n14, 265
compadre (Latino concept of interconnectedness), 168
compost, 27, 178–80, 259n7
conservation, 139, 175, 192, 271
consumers and consumerism, 1–2, 59, 71, 103, 174–75, 222, 238
Cook, Martha E., 21

Coonrod world renewal ceremony (Coonrod Flat, California), 78
corn: as ancestors' gift to descendants, 71, 225; as ancestor to humans, 200, 210; blue, 228; changes in, 150; Corn Mother (Mohawk), 227; in cosmologies, 211, 213, 227, 236; cultural identification with, 3, 191; domesticated, 21; GMO, 191; illustration, *201*; as life model, 200; in Native American cosmologies, 211, 213, 227–28; "Onëö (Word for Corn in Seneca)," 210–13; overplanted, 35; red (Mohawk red), 225; revitalized, 225; ritual planting of, 225–26; spirituality in, 210–12; white (Iroquois), 223; white (Mohawk), 228; white (Seneca), 210–11, 213, 264
corporations and corporate rights, 63, 128, 131–32, 143, 174–75, 269
cosmogenealogies, 4, 227
cosmologies: ancient vis-à-vis modern, 145, 147; birds in, 163–64, 259nn4–5; corn in, 211, 213, 227, 236; East Indian, 188; Māori, 109, 232–33; modern, 149, 153; Native-American, 211, 213, 227–28; of the Winnemem Wintu, 73–75, 77–79. *See also* Winnemem Wintu Tribe
Coulter, Burley, 40
creation legends, 151, 208, 226–28, 232–33
Creator: of the Anishinaabeg, 143–44; in colonialism, 52; Earth as, 146–47, 149; food as, 190; food provided by, 223; instructions from, 128; invoked, 53; messages from, 250; of the Winnemem Wintu, 73–75, 77–79
Cree, Mary, 219
Cultural Conservancy (San Francisco Indigenous-led organization), 223, 255, 261n5, 269, 270. *See also* GOTW Native Youth Guardians of the Waters Program) (San Francisco Bay Area)
Cummings, Claire, 261n5
Cummings, Katherine Kassouf, 90–94, 264

Dahl, Michael, 1, 5, 126–28, 186–87
Dakota Nation/tribes, 3, 230

Index   277

dance: ancestry honored in, 3; of Big Head spirits, 83; and the GOTW program, 209; and grieving for nature out of balance, 92–93; knowledge found in, 199; learned by Native youth, 209; of the Pottawatomi in pre-Chicago, 90; as ritual healing, 94; skiing as, 124, 128–29; slow, of the land, 24; war dance, 76, 269

Darwin, Charles, 153

Darwin's Worms (Phillips), 259n7

Day of the Dead celebration, 62

decay and decomposition, 29, 89, 91–92, 178–81, 183, 231

decomposition. *See* decay and decomposition

Delaware Nation/tribes, 200

Department of Resources (California), 82

descendants: as ancestors, 147; ancestors providing for, 56–59, 141, 143, 172, 183, 197–99, 215–16, 225; ancestors without, 126, 222; as ascendants, 183–84; corn as ancestors' gift to, 225; indebted to ancestors, 187–88; of other-than-human ancestors, 5, 31–32, 56, 155, 185, 189–90; seeds as, 200; of the Seventh Fire, 220–21; of white farmers, 66; of the Winnemem Wintu, 83. *See also* "To Future Kin" (Calvert poem)

Detroit, 7, 112–14, 267–68

Diamond, Jared, 64

Dillard, Annie, 237

Diné (Navajo) homelands and lineage, 266

DNA (biological), 3–4, 58, 60, 61–62, 206, 234

DNA (Descendants-N-Ancestors), 3, 220

Dominion of the Dead, The (Harrison), 104

Doyle, Brian, 70

drought, 21, 74, 83, 181, 200

drumming, 93, 231, 247–48

Drums of Winter (film), 101

Duke, James B., 34

Dungy, Camille T., 110–16, 264

Earth: age of, 161; ascendants on, 183; cognizance of, 146; as Creator, 3, 146–47, 149; despoilers of, 59; finite resources of, 222; human cognizance of, 140, 150; human friends of, 166–71; human interdependence with, 209; humans returning to, 3, 29, 31, 215–17, 231; life emerging on, 130; as living integrated system, 131, 169; and LUCA, 130; in Māori cosmologies, 96, 109; and mourning, 216–17; as place, 192; as protector, 146; Pueblo context of, 133; redemption of, 146; responsibilities toward, 192; as a Spell, 107; uncolonized spirit of, 221; warnings about destruction of, 69–70; you, as steward of, 194. *See also* biodiversity; Mother Earth; "Nourishing" (White essay); seeds

Earth Guardians (international conservation organization), 207, 223

Earth Island Institute (Berkeley, California), 268

"Earthly," 5, 157–94; epigraph, 155; essays, 159–65, 166–81; interview, 185–93; poems, 157–58, 194

Earth mother, 96, 217, 232–33. *See also* Mother Earth

ecological amnesia, 143

ecological ancestry, 179

ecological cognizance, 189

ecological communities/families, 5, 167–68

ecological-cultural interconnections, 37, 127, 148, 208

ecological death-facing, 179

ecological health, 127, 148, 208, 266

ecology: crises in, 216; ethnoecology, 166, 269; feminist, 268; in the GOTW program, 207; humans impacting, 2, 125, 179; inheritance ethos vs, 179; "Kinetic Ecology," 167; mindsets of, 189; mourning over ecological loss, 178–79; multicultural knowledge applied to, 216; queer, 268; responsibilities toward, 35; restoration of, 267; as sacred, 238. *See also* climate change/disruption; land; water

ecospheres, 22, 153, 265, 270

ecosystems: ancestors of modern land plants as vital to, 21; ancient, 138; anthropocen

ecosystems (*cont.*)
trism vs, 185; collapsing, 69; and fire, 92; health in, 127, 148, 208, 266; human body as ecosystem, 186; Indigenous cultures absorbing knowledge of, 132; vs industrial mindsets, 148; natural processes of, 132, 148; and overfishing, 244–45; as places, 134; planetary commons of, 133; restoration of, 34; restored, 138; rice-based, 127; and snow, 124; wildness preservation among, 148

Edelman, Lee, 179–80

Egypt, 240–42

Eighth Fire, 4, 221, 261n4

Elder, Sarah, 101

"Embedded," 4, 13–45; epigraph, 11; essays, 16–19, 20–24, 25–32; interview, 33–43; poems, 13–15, 44–45

Emery lineage, 28

empathy, 58, 201, 221

England, 61, 63–64, 152

Enote, Jim, 212

Entrada, 146, 166

environment: activism and advocacy, 22–23, 77, 139, 143–44, 260n4, 264, 268; biodiversity creating stability in, 135; and bioregionalism, 135; consciousness change required for healing, 249; disconnection from, 235; environmental injustice, 248; environmental justice, 181, 214, 265–66; food sources adapting to changes in, 211; and human cultural variation, 152; human relationships with, 212; human unity with, 235, 241; Māori ancestors as guardians of, 234, 238; nurturing of, as self-preservation, 137; as a place of learning, 20; worsening conditions of, 248–49. *See also* Communities for a Better Environment (CBE) (Los Angeles County); ecology; ecosystems

Eocene epoch, 140

Esselen Nation, 222

ethnicity, 134, 266

ethnoecology, 166, 269

Everhart lineage, 28

evolution: as adaptation vs competition, 131; of ascendants, 184; of cognition, 130–31, 145; cultural identification with, 136; evolutionary responsibility, 66; failing, 138; human, from tree-dwellers, 21; from hunting/gathering to farming, 64; indeterminacy of, 232; of knowledge, 145; and Māori genealogical-based relationships, 233; personal, 174; in recent millennia, 136; and skin color, 65–66; spiritual development in, 135

Ewell, Jack, 148

exceptionalism, 152

extraction and extractivism, 69, 146, 221

Fairbanks Climate Action Coalition (Alaska), 266, 270

family histories, 3, 25–32, 110, 113–15, 139. *See also* genealogies

farms and farming: by ancestors, 64–65, 166; ancestral practices of, in India, 188; and the Berry Center, 43; Berry on, 34–36, 42; Black-owned farms, 58; corn farming, 210–11, 225–31; depression-era, 34; droughts threatening, 21; in Europe, 64; by Indigenous youths, 224; industrial, 82; by Native Americans, 225–31; natural processes returning to, 148; in the Neolithic Period, 64–65; and skin pigmentation, 64–65; stolen, 58; suicide among farmers, 188, 190. *See also* agriculture

Father Sky, 208, 221, 232

Ferrer, Antonio, 202–3

Fifth Fire, 220

fire, 4, 90–92, 94, 125. *See also* "Ancestor of Fire" (Abeyta essay); Eighth Fire; Seventh Fire; Sixth Fire

First Nations people (Alaskans), 108, 147, 187, 248

first people (Native Americans), 220, 224

flax, 95–98

Florissant Fossil Beds National Monument (Colorado), 139

Foote, Jim, 42

Forbes, Peter, 172–76, 264
forests: and ancient plant life, 175; of Aotearoa, New Zealand, 235–37; burning beyond regeneration, 92; decay as nourishment for, 178; decay nourishing, 183; destroyed, 69; forest fires, 125; Forest God of the Māori, 96, 104; and Grace Trail, 177–78, 181; rainforest experiments, 148; and snowpack reduction, 140
Forest Service (US), 80–81
"Forgiveness?" (Gibney poem), 49–50
forgiveness/forgiving, 51, 160, 202
"Formidable" (Moore essay), 68–72
fossil fuels, 69–71, 82, 142–43, 147–48, 186
fracking, 82, 268
French Expedition against the Haudenosaunee (1687), 213

Galiano Island (Canada), 177
Gandhi, Mahatma, 127, 189
Gansworth, Kristi Leora, 121–23, 264
genealogies, 3–4, 28, 96, 222, 227–28, 232–33. *See also* cosmogenealogies; family histories
genocide, 152, 166, 243–48
Geography of Hope (Black Mountain Circle) (Pt. Reyes, California), 255
Geological Survey (US), 134
geologic epochs/periods: Anthropocene, 178, 180; Eocene, 140; and human evolution, 64; Ice Age, 64, 132; Mesolithic, 62, 64; Neolithic, 62, 64; Ordovician, 161; Pleistocene, 62, 132, 136, 161; Quaternary, 161; Silurian, 161; Triassic, 161
ghosts, 6, 218, 222, 261n17
Gibney, Shannon, 49–50, 264–65
Gilbert, Jack, 112
glaciers, 26–27, 31, 92, 178, 190
globalization, 2, 222, 269
global warming, 43, 69, 71, 125, 150
Goeman, Mishuana, 216
GOTW (Native Youth Guardians of the Waters Program) (San Francisco Bay Area), 200–209, 223

Gould, Corrina, 261n16
Grace Trail (Galiano Island, Canada), 177–78, 181
Graham, Linda E., 20
Grand Canyon (Arizona), 139
Grande, Sandy, 215–16
Great Blue Heron, 24
Great Depression, 34
"Great Granddaddy" (Coleman poem), 13–15
Great Plains, 21
Green Party, 267
Green Path (East Indian eco-initiative), 223
Greenwood (Mississippi), 13
grieving and mourning, 178–79; as cultural process, 18; functions of, 222; as healing, 93, 223; intergenerational, 220; as a necessity, 144; rituals of, 92–94; for snowpack loss, 125; and the soil, 216–17
ground. *See* ancestral grounds/lands/soil; soil
"Grounded" (Krug essay), 20–24
Guardians of the Earth. *See* Earth Guardians (international conservation organization)
Guardians of the Waters. *See* GOTW (Native Youth Guardians of the Waters Program) (San Francisco Bay Area)
Gucker lineage, 28
Guerrero, Aurora, 214, 260n1
Guttierez, Oscar, 214–18, 265
guns and weapon technology, 65–66, 84
Gwich'in First Nations people (Alaska), 108

Halberstam, J. Jack, 180
Hall, Edward T., 134–35
Hall lineage, 28
Haraway, Donna, 178–79, 180
Harjo, Maya, 200
Harrison, Robert Pogue, 103
Haudenosaunee Confederacy, 211, 213, 264
Hausdoerffer, Atalaya, 124–25
Hausdoerffer, John, 1–5, 119, 124–29, 145–53, 185–93, 265
Hausdoerffer, Sol, 124–25
Hawaiians, 4, 234, 249

Headwaters Conference (Western Colorado University), 255

"Healing," 4, 89–117; epigraph, 87; essays, 90–109; interview, 110–16; poems, 89, 117

healing: ancestors promoting, 101; from colonization, 4, 87, 220–23, 225; damaged lands, 138; Earth promoting, 54; GOTW promoting, 207–8; grieving as, 93, 223; land relationships promoting, 173; people committed to, 104; rifts between Indigenous people and colonizers, 223; rituals promoting, 94; songs of, 231; through technology, 98. *See also* medicine (natural); medicine (ritual/traditional)

Hecht, Brooke Parry: and the Center for Humans and Nature, 265; in Merculieff interview, 240–42, 244–51; in Sisk interview, 73–75, 78, 80, 82–83

Hecht, Tara, 250

Hemmings, Sally, 59

Henderson, Sakej, 261n1

Herron, Elizabeth Carothers, 157–58, 265

Hinduism, 187, 189

Hingston lineage, 28

Hinojosa, Mateo, 211

historical amnesia, 143

Hitler, Adolph, 185

Hogan, Linda, 3, 252–53, 265

Honor the Earth program (Callaway, Minnesota), 267

Hopi Independent Nation, 133–34, 260n5

"How to Be Better Ancestors" (LaDuke essay), 142–44

humans: age of, as a species, 132, 136, 161; allowing other-than-humans to flourish, 185; Black people considered to be nonhuman, 57; centrism of, 136; compassion/kindness as legacy of, 154; corporate rights vs, 269; cultural variations among, 152; ecological health benefitting, 148; human body as ecosystem, 186; as humus, 20, 183; Indigenous mindfulness in, 130–37; migrating to the Western Hemisphere, 132; and the more-than-human world, 176; out of balance with Nature, 4; overpopulation of, 92; as part of nature, 135, 137; as Pleistocene hunter-gatherers, 132; protective instincts of, 71; race in, 62; the real human being (Unangan concept), 242, 249; survival of human communities, 1. *See also*; creation legends; fire; "Inheritance, An" (interview); other-than-human beings

humus, 20, 183–84, 259n7

"Humus" (Sandilands essay), 177–81

hunter-gatherers, 62–64, 132, 147

hunters and hunting, 63, 219

Huntington Park (California), 214–15

Iban society (Borneo), 163

Ibarra, Jessica Garcia, 203–5, **204**

Igadagik (son of Unangan chief), 243–44

Illinois, 90–92

India, 186–89, 191, 193, 269

Indiana, 57

Indian Canyon (California), 209

Indian People Organizing for Change (IPOC), 261n15

Indian People Organizing for Change (San Francisco Bay Area), 261n16

Indians: American Indian Studies, 170, 269; bounties on, 76; and Columbus, 187; ecological sophistication of, 167–68, 216, 227, 238; Pueblo Indians, 133; uprisings of, 142, 246; urban Indians, 219. *See also* Indigenous people; Native Americans

Indigenous Impact Rapid Response Initiative (Pueblo Alliance; Seeding Sovereignty), 271

Indigenous people: ancestry revered by, 3; and bioregionalism, 135–36; ecological sophistication of, 167–68, 216; expansiveness of, 78, 98; genocide perpetrated against, 152, 166, 243–48; of the Kansas plains, 148; knowledge transferred by, 166; leadership of, 77; mindfulness of, 130–37, 146;

murdered, 76, 95; non-Native ecologists in agreement with, 169; oppression of, 52, 95, 166, 220, 248–49, 269; othering of, 103; places as ancestors of, 134; suicide among, 54; water as sacred to, 77. *See also* GOTW (Native Youth Guardians of the Waters Program) (San Francisco Bay Area); Seeding Sovereignty (Indigenous woman-led resistance organization); *and specific cultures (e.g., Abenaki Tribe/Nation)*

"Inheritance, An" (interview), 33–43

interconnectedness, 96, 112, 133–34, 167–69, 235

International Crane Foundation (Wisconsin), 268

Interstellar (film), 206

"Interwoven," 4–5, 121–54; epigraph, 119; essays, 124–29, 130–37, 138–39; interview, 145–53; poems, 121–23, 154

investing vs acquiring, 206

Ireland, 172, 234

"I Want the Earth to Know Me as a Friend" (Salmón), 166–71

iwígara (Rarámuri concept of interconnectedness), 168–69

Jackson, Wes, 42, 144, 145–53, 265

Jacobs, Anna, 229–30

Japan, 132

Jeffers, Robinson, 223, 261n6

Jewish people, 142, 149

Jicarilla Apache Nation, 54

Johnson, Princess Daazhraii, 100–109, 266

Johnston, Lyla June, 197–98, 266

Jones, Florence, 268

Journey for Our Existence Movement (prayer walk in Diné homelands), 266

Journey of the Universe project, 150

Kakugawa, Frances H., 44–45, 84–85, 117, 154, 194, 266

Kamchatka, 240

Kamerling, Len, 101

Kansas, 20–21, 42, 144

Kanza Nation, 148

Kauai (Hawaii), 249

Kaur, Valarie, 106

Kaw Nation, 148

#KeepItIntheGround! working group, 270

Kentucky, 35, 43, 263

Kentucky River, 35

Kenya, 192

Kernza genome, 147, 150

Kimmerer, Robin Wall, 2, 182–84, 215–16, 224, 266

"Kincentric Ecology: Indigenous Perceptions of Human-Nature Relationships" (Salmón"), 167

Kinder Morgan Pipeline project, 260n4

Kitahtëne peaks (Vermont), 25, 27–28, 30

Klamath River, 82

"(Korowai, A) For When You Are Lost" (Sweeney essay), 95–99

Kropotkin, Peter, 136

Kule Loklo (Point Reyes Miwok village recreation), 209

Kunuk, Zacharias, 262n2

LaDuke, Betty, 142

LaDuke, Winona, 5, 125, 142–44, 186–87, 190, 267

Laguna Pueblo (New Mexico), 54, 271

Lake Michigan, 159–65

lakes: as ancestors of humans, 219; as ancestors to humans, 32; justice demanded for, 142; lights moving on, in New Zealand, 236–37; as restoration sites, 209; rice harvesting on, 127–28, 142; Round Lake, 126, 142; Solstice Lake, 25–27, 31–32

La Montaña (Northridge earthquake rubble pile), 215, 260n2

land: as ancestor to humans, 166, 167, 255; ancestral homelands, 25, 80, 90, 93, 95, 125; as ancestry barometer, 119, 128; ancestry found in, 217; as community, 133; contaminated, 43, 84, 145, 150, 160, 214,

land (*cont.*)
220; extractivism vs, 221; farmers losing connection with, 190; Indigenous people's connections to, 134; introduction to, required, 167; as memory, 216; "Of Land and Legacy," 56–60; stolen, 54, 58, 83, 174, 220; and US nationhood, 173. *See also* ancestral grounds/lands/soil

"Landing" (Guttierez essay), 214–18

Land Institute (Salina, Kansas), 144, 146–47, 153, 265, 270

LaPena, Sage, 4

"LEAF" (Herron poem), 157–58

"Learning a Dead Language" (Merwin), 107

Lenapewihittuk River, 26–31

Leopold, Aldo (husband), 138–41

Leopold, Estella (daughter), 138–41, 267

Leopold, Estella (wife), 138–39

Leopold, Luna (son), 138–39

Leopold, Nina (daughter), 139

Leopold, Starker (son), 138–39

Levant, 132

Lewis, C. S., 38

LGBTQ communities. *See* queer communities

Library of Congress, 152

"Light" (Spiller and Wolfgramm essay), 232–39

"Light by Chellie" (Spiller), 235–38

"Light by Rachel" (Wolfgramm), 234–35

light-skinned people, 195. *See also* white people

Lindsey, Elizabeth, 235–36

Little Bluestem bunchgrass. *See* Bluestem roots

Loeffler, Jack, 130–37, 267

Lopez, Camillus, 136–37

Lorde, Audre, 112

Los Angeles, 200, 214, 217, 265

"Lost in the Milky Way" (Hogan poem), 252–53

Love, David, 140

LUCA (last universal common ancestor of all life), 130

Lucich, Luna, 206

Luisi, Per Luigi, 131

Lukacs, John, 38

Luna (feminine archetype), 139

Luna, Solomon, 139

Luna lineage, 139

"Luna's Free Style Writing" (Lucich essay), 206

Lunsford, Emmet, 57–58

Lunsford, Frances, 59

Lunsford, Lindsey, 56–60, 267

Lyons, Oren, 211

Machu Picchu (Peru), 234

Mad River Valley, 264

magic, 60, 75, 164, 250–51

mana (Hawaiian and Māori life spirit), 4

manifest destiny, 66

manitou (Anishinaabeg life spirit), 4

Māori people of New Zealand: ancestor-figures of importance, named, 234–35, 237; ancestors (*Tupuna*) of, 96; and the Aotearoa mountains, 107, 234–35; attacked, 218, 248; colonization of, 107; creation legends of, 232–33; flax in the culture of, 95–98; genealogical narrations (*whakapapa*) of, 3, 232–33; guardians/protectors of, 208, 216, 266; *korowai* cloaks of, 3–4, 95–99, 232–39; light in cosmologies of, 233–34, 236–37; Mother Earth (*Papatūānuku*) of, 96, 232–33; resistance/resilience of, 106, 215; Sky father (*Ranginui*) of, 96; Spiller's advocacy for, 269–71; spirituality of, 232–33, 238; Te Whakatohea tribe of, 235; vibrational call (karanga) of, 234

Mapuche tribes (Argentina), 241

Mason-Dixon line, 57

Maturana, Humberto, 130

May, Jamaal, 6–9, 267

Maya people (South America), 205

McCloud River, 76, 79, 260n6, 269

McLeod, Toby, 73, 76–77, 82, 268

Meadow Voles, 23

medicine (natural), 14, 121, 143, 182, 212, 224
medicine (ritual/traditional), 52, 54, 100, 202, 209, 266
Meine, Curt, 138–41, 268
Merculieff, Ilarion, 240–51, 268
Merwin, W.S., 107–8
Mesolithic era, 62
Métis Indigenous people/language of North America and Canada, 3, 223, 269
Mexico, 140–41, 168–69, 205, 234
Michif (Cree language), 219
Michigan, 159. *See also* Detroit
Michoacàn (Mexico), 205
Midewiwin Lodge (Anishinaabe), 126
Miller lineage, 28
Minnesota, 1, 125–27, 142
Miranda, Deborah, 217, 222
misogyny, 63
Mississippi, 13–14
Miwok Indians/Tribe, 1, 209
Mohawk Tribe, 225–26, 228
Mongolia, 240, 242
Moore, Kathleen Dean, 68–72, 268
more-than-human beings, 4, 23, 176, 255, 270
Mosquita y Mari (film), 214, 260n1
Mother Earth: attacked, 218, 248; in creation legends, 228, 232; and the decolonized spirit, 221; guardians/protectors of, 208, 216, 266; resistance/resilience of, 106, 215; as slave to exploitation, 143–44; as source of human life, 3, 104. *See also* Earth mother; Original Woman (Mohawk creation figure); PachaMama (Andes Mountain peoples' fertility goddess)
Mountain Resilience Coalition (Colorado), 265
mountains: of Alaska, 104, 242–43; as ancestors of humans, 203–5, 234; ancient energy inhabiting, 105; Cinnamon Mountain, 125; Cold Mountain, 26–30; Elk Mountains, 124–25; forest fire fuel accumulation on, 92; human displacement in, 186; as kin, 234; Kitahtënē peaks, 25, 27–28, 30; light from, 237; of Machu Picchu, 234; mountain springs, 3; of New Zealand, 96–97, 234, 237; Olympic Mountains, 177; Purple Mountain, 125; Rocky Mountains, 54; as sacred places/sites, 80–81, 83; salmon runs in, 75–76, 79; San Juan Mountains, 177; snow hunting in, 124; Solstice Mountain, 25, 31–32; Turtle Mountains, 219, 224; Ute Mountain, 260n5; in Winnemem Wintu creation legends, 73–74, 77. *See also* La Montaña (Northridge earthquake rubble pile)
Mount Shasta (Buliyum Puyuk), 269
Mount Shasta (California), 73, 80–83, 209, 260n6, 268. *See also* Shasta Dam (California)
Mount St. Helens (Washington), 139, 183
mourning. *See* grieving and mourning
"Moving with the Rhythm of Life" (Cummings essay), 90–94
Mumbai (India), 187
Muñoz, Jose Esteban, 261n15
Muscogee Creek Nation, 200
"My Home / It's Called the Darkest Wild" (Prentiss essay), 25–32
"Mystery of the Alaskan Mummies, The" (documentary), 246–47

National Academy of Sciences (US), 138
National Outdoor Leadership School (Lander, Wyoming), 174
Native Americans: as ancestors, 146; bounties on, 76; cosmologies of, 211, 213, 227; creation legends of, 151; ecological sophistication of, 167–68, 216, 227, 238; Europeans' decimation of, 132; as the first people of the Americas, 220, 224, 248; foodways of, 208–9, 264; and fossil-fuel use, 151–52; habitat diversity of, 132; lands stolen from, 54, 66, 83, 174, 220; and manifest destiny, 66; place-sense of, 170–71; in a sun-powered world, 146; uprisings of, 142, 246; US population of, 133; in the Western

Native Americans (*cont.*)
Hemisphere, 132; youth identifying as, 209. *See also* Indians; Indigenous people; and specific cultures (*e.g., Abenaki Tribe/Nation*)
Native Youth Guardians of the Waters Program (GOTW) (San Francisco Bay Area), 200–209, 223
nature, 235; anthropomorphic exclusion of, 191; biodiversity creating stability in, 135; in charge, 198; connections with ancestors through, 205; destruction of, as destruction of the future, 191; ecological principles active in, 132; GOTW's activities on behalf of, 208; guardians/protectors of, 234; human indebtedness to, 187; humans as part of, 135; humans' loss of place in, 136–37; Indigenous people living in harmony with, 132; Māori gods of, 233; nature/nurture duality, 115; original elements of, 101; source, as its own, 130; and the Tao, 159–61, 163; war against, 189; wisdom of, evolved, 130; working relationships with, 173–74
Navajo Nation, 223, 260n5, 266
Navdanya (women/farm/ecology-based organization) (New Delhi, India), 188, 192
Nayarit (Mexico), 140
Nebraska, 21–23
Nelson, Melissa K., 3, 219–24, 269
Neolithic period, 62, 64
New England, 7, 264
New Hampshire, 128–29
New Mexico, 102–3, 139, 211–12, 266, 270
New York City, 142
New York State, 210–11, 213, 219, 266
New Zealand, 76, 107, 234–35, 260n6, 270. *See also* Māori people of New Zealand
nitrogen, 150, 178
Nobel Prize, 149–50, 154
Noptipom Wintu Tribe, 4
North Dakota, 219, 224
Northridge earthquake, 215, 260n2
"Nourishing" (White essay), 225–31

nuclear holocaust, 150
nuclear waste, 139
nuclear weapons, 145–46

Oakland (California), 214–15, 261n14
Obama, Barack, 63
oceans: artesian spring water feeding into, 79; contaminated/poisoned/polluted, 69, 84, 92; and the GOTW program, 209; life originating in, 130, 149; and Salmon cycles, 75, 78; Tao of, 160–61
Odawa Indians, 4
"Of Land and Legacy" (Lunsford essay), 56–60
Ohlone Tribe, 209, 222
Ojibwe People, 4
Oklahoma, 161, 200, 212
Ollero clan (Jicarilla Apache), 54
Omaha (Umóⁿhoⁿ íe tʰe) language, 21
Omaha (Umóⁿhoⁿ) Tribe, 21–22, 24
"Omoiyare" (Kakugawa poem), 154
Oneness versus the One Percent (Shiva), 190
"Onëö (Word for Corn in Seneca)" (Bray essay), 210–13
Onondaga Nation, 211–12
Ordovician period, 161
Organic Machine Nation, 267
Original Woman (Mohawk creation figure), 226–27
Osage Nation, 161–62
Otero Lineage, 139
other-than-human beings: acknowledgement of, 216; as ancestors or descendants, 5, 31–32, 56, 155, 188–89; human relationships with, 22–23, 104, 270; humans sharing ecosystems with, 133; lands as, 90, 166; leaves as, 157–58; as teachers, 103–4
Ouray Reservation, 260n5
oxygen, 77, 149–50

PachaMama (Andes Mountain peoples' fertility goddess), 204
Panoho, John, 270

People of the Fire. *See* Pot(t)awatomi Nations
Peru, 234
Peters lineage, 28
Phillips, Adam, 259n7
Piailug, Mau, 236–37
Pilgrim at Tinker Creek (Dillard), 237
Pilgrims and Tourists (film), 82
place-sense: agricultural, 34–36, 225; and ancestry, 11, 128, 181, 193; of the Anishinaabeg, 142–44; and bioregionalism, 135–36; of cultural disaster, 243–47; as hope, 107; of the Hopi, 134; of the Māori, 96, 233; misguided, 81; of Native Americans, 170–71; of place-based people, 3; and place-based values, 270; and self-identity, 170; and Standing Rock, 143; as understanding places' messages, 169–71; as well-being, 90. *See also* sacred places
Pleistocene epoch, 62, 132, 136
Pollen Path (Navajo blessing), 223
pollution. *See* air pollution; land; oceans; water pollution
polyculture(s), 147–48, 150, 263, 265
Pomo Indians, 209
Posey, Darrell, 261n5
Pot(t)awatomi Nations, 4, 22, 90–92, 266
prairies: fires nurturing, 91–94; functional groups in, 148; as homelands, 146; of Illinois, 91–92; Indigenous people on, 20, 91–92, 147; of Kansas, 146; and The Land Institute, 146; of Nebraska, 20, 22–23; prairie roots, 22; the wild and the domestic in, 148
prayers: in Alaskan all-women ceremonies, 250; alternatives to, 18; of ancestors, 229; to ancestors, 74, 79; of ancient Hawaiians, 236; of the Anishinaabeg, 142; coming to life, 54; for corn harvest, 226–31; of the Creator, 79; to the Creator, 250; to the Fire Spirit, 74; of Jessica Garcia, 204; of the Māori, 96; misguided, 80; of the Mohawk, 226–31; mountains as places for, 74–75, 83; for peace, 248; prayer journeys through Diné homelands, 266; to the right gods, 14; for Salmon, 75–76, 78; of the Winnemem Wintu, 73–83

Prentiss, Grandy, 25–26, 28–30
Prentiss, Sarah Eve, 25–32
Prentiss, Sean, 25–32, 269
Prentiss, Winter Eve, 25–32
Prentiss lineage, 28
Pribilof Islands (Alaska), 243–45
"Promises, Promises" (Kakugawa poem), 84–85
Pt. Reyes National Seashore (California), 209
Puc, Nance, 205, **205**
Pueblo Nations, 54, 133, 260n5, 269, 271
Pyne, Stephen J., 91–92

Quapaw Tribe of Indians, 200
Quaternary period, 161
queer communities: Chican@s in, 202; and Eighth Fire people, 223; Mexican Americans in, 205; in *Mosquita y Mari*, 260n1; Native Americans in, 180; theorists of, 179–80, 261n16, 268

racism, 57, 59, 63, 66, 71, 220–21
rain, 31, 35, 79, 128, 212
Raine, Kathleen, 38
rainforests, 148
rape, 58
Rarámuri Indigenous people (Mexico), 168
"Reading Records with Estella Leopold" (Meine essay), 138–41
Reagan, Ronald, 53
"Reckoning," 4, 49–85; epigraph, 47; essays, 51–55, 56–60, 61–67, 68–72; interview, 73–83; poems, 49–50, 84–85
Reconquista, 201
Red Road (Native American spiritual ways), 220, 223
regeneration: ancestors bringing about, 3–4, 23, 229; and cognition, 131, 136; of corn seeds, 211; of ecosystems, 133; fires destroying processes of, 92; Mohawk

regeneration (*cont.*)
  prayers of, 229; regenerative farms, 264; as requisite for life, 189; of soil, 189; and soil quality, 182–84. *See also* restoration
Regeneration Festival (New Mexico), 266
"Regenerative" (Nelson essay), 219–24
religion: and ancestral love, 197; and colonialism, 52; forgiveness in, 51; Hinduism, 187, 189; Indigenous, 74; limitations of, 42; and Mount Shasta, 80–81; promises of conditions beyond, 84; represented in the Library of Congress, 152; and slavery, 52, 65
reproductive futurism, 179
Research Foundation for Science and Technology (India), 188
resilience: of corn, 213, 230; dance expressing, 231; good soil enabling, 182; of hurricanes, 127; intergenerational, 231; of the Māori, 98–99; of Mother Earth, 104; Mountain Resilience Coalition, 265; of places, 169; Resilience Studies Consortium, 265; of seeds, 229; spiritual, 190; against traumas of colonialism, 220; as a way of life, 188, 202
Resilience Studies Consortium (educational network), 265
resistance: and ancestry, 54–55, 103, 188–89, 202; by the Canadian Sioux, 260n4; against the Kinder Morgan pipeline project, 260n4; by Mother Earth, 215; and seeds, 229; against the Shasta Dam, 269; on the Standing Rock Sioux Reservation, 77–78, 143; of trees, 182; against voluntary infertilization, 60
restoration: of balance in life, 103; of the Earth, 146, 194, 208–9; ecology of, 267; of human relationships, 23; of Salmon runs, 269; of Turtle Island, 220. *See also* regeneration
"Restoring Indigenous Mindfulness within the Commons of Human Consciousness" (Loeffler essay), 130–37

rice and rice farming, 124, 127–29, 142–43, 187, 190
Rio Gavilan (Mexico), 141
rituals: as ambiguous, 189; of ancestors, 18; ancestors honored in, 3; community created in, 94, 209; of introduction, 169; of the Māori, 168; of mourning, 93–94, 98–99; re-entry from, 99; ritual space, 167; of storytelling, 37; unifying qualities in, 87, 98
rivers: as ancestors, 104, 234; ancient energies inhabiting, 105; Chicago River, 131; Colorado River, 124, 139; Columbia River, 139; dams destroying, 75, 83, 139, 209; East River, 124; Gunnison River, 124; Kentucky River, 35; Klamath River, 82; Lenapewihituk River, 25–31; Los Angeles River, 214; Manawatu River, 109; McCloud River, 76, 79, 260n6, 269; polluted, 82; as restoration sites, 209; Rio Gavilan, 141; and riverbeds, 77; Rocky River, 68–69; Salmon runs in, 75–79; Smoky Hill River, 21–22; songs of, 141; Tao of, 160–61; Wisconsin River, 131, 141; Yukon River, 104
Roadrunner clan (Apache), 54
Rocky Mountains, 54
Roske-Martinez, Xiuhtezcatl, 223
Round Lake (Minnesota), 142
Run4Salmon ceremony, 75–76
Russia, 241, 243–46

Sacred Land Film Project (Earth Island Institute) (Berkeley California), 268
sacred places: around the Shasta Dam, 75, 83; ceremonial introduction to, 167–69; of the Gwich'in Nation, 108; misguided reasons for visiting, 80–81, 237–38; in New Zealand, 235; of the Unangan islands and people, 244–45
salmon (fish), 3, 75–76, 78–79, 260n6, 269
Salmon (spirit), 73–74. *See also* Run4Salmon ceremony
Salmón, Enrique, 166–71, 217, 269
Salt, Waldo, 102

Sand County Almanac, A (Aldo Leopold), 138
Sandilands, Catriona, 2, 177–81, 268
San Francisco Bay Area, 78, 170, 219, 223. *See also* GOTW (Native Youth Guardians of the Waters Program) (San Francisco Bay Area); Shellmounds (San Francisco Bay Area)
Santa Clara Pueblo (New Mexico), 133
Satyagraha (adherence to personal values), 127
Sauk Prairie Conservation Alliance (Wisconsin), 268
Scott, Donald M., 66
sea levels, 69, 92, 124
Searles lineage, 28
Seeding Sovereignty (Indigenous woman-led resistance organization), 271
seeds: as ancestors and descendants, 200; ancestral seeds, 4, 193, 213, 224, 225–28, 229–30; carried for the future, 197; cultivation of, 21; *harakeke* (flax seeds), 95, 97–99; maple seeds, 182–84; polycultures of, 148; seed banks, 184, 192–93; and seedkeepers, 228–29, 270; songs of, 229–31; terminator seeds, 189. *See also* corn; GOTW (Native Youth Guardians of the Waters Program) (San Francisco Bay Area); "Onëö (Word for Corn in Seneca)" (Bray essay)
"Seeds" (Wagenberg/Guardians of the Waters essay), 200–209
Seminole Nation, 200
Seneca Nation of Indians, 210, 264
Seventh Fire: and ancestors-in-practice, 5; the Anishinaabe as people of, 187; the Anishinaabeg as people of, 4; descendants of, 220; leadership of, 255; and the Light-skinned Race, 195; new people emerging in the time of, 195
"Seventh Fire," 5, 197–253; epigraph, 195; essays, 200–209, 210–13, 214–18, 219–24, 225–31, 232–39; interview, 240–51; poems, 197–99, 252–53
Seventh Generation principle, 187
sexism, 221

sexual violence, 53, 58, 63
Shasta Dam (California), 75–76, 83, 260n6, 269. *See also* Mount Shasta (California)
Shawnee Tribe, 200
Shellmounds (San Francisco Bay Area), 218, 261n15
Shepard, Paul, 132
Sherman, Alexie, 218
Shiva, Vandana, 155, 185–93, 269
Siberia, 240, 242
Silurian period, 136, 161
Simpson, Leanne, 221
Sioux Nations. *See* Canadian Sioux; Dakota Nation/tribes; Sitting Bull; Standing Rock Sioux Tribe and Reservation
Sisk, Caleen, 4, 73–83, 269
"Sister's Stories" (Wise essay), 51–55
Sitting Bull, 144, 230
Sixth Fire, 220, 221
skiing, 124, 128–29
Skip in the Day Family, 142
slavery: absent from Library of Congress murals, 152; in *Amistad*, 62, 143; of ancestors, 59; casual accounts of, 32; chattel slavery, 152; and Christianity, 52, 65; and gun control, 65; as legal commerce, 173; Mother Earth enslaved to exploitation, 143; of Native Alaskans by Russia, 243–44, 246; virtual, 59. *See also* "Of Land and Legacy" (Lunsford essay)
Slow Food, 187
Smoky Hill River (Kansas), 21–22
snow, 25–26, 28, 31, 79, 124–25, 129
Snyder, Gary, 135–36
soil: and ancestry, 4, 20, 23, 30, 32, 128, 183–84; bioregional, 135; "Building Good Soil," 182–84; communities of, 22–23; cycles, 28; ecologies protecting, 148; enrichment, 128; erosion, 145, 150, 178; fire enriching, 91; humans returning to, 227; and humus, 178, 180–81; lost, 21, 160; modern, 21; and mourning, 216–17; nourishment of, 175; of prairies, 22; preservation,

soil (*cont.*)
    in prairies, 148; regeneration of, 189; and solar energy, 153; vegetation creating, 178. *See also* ancestral grounds/lands/soil
solar power, 151
Soliev (Russian officer), 246
Solstice Lake (Vermont), 25–27, 31–32
Solstice Mountain, 25, 31–32
songs and singing: of ancestors, 100, 107; ancestry honored in, 3; of birds, 44–45, 98, 109, 234; of the Cold Mountain poet, 30, 32; for a dying grandmother, 53; earned, 55; to the Earth, 169, 205; and the GOTW program, 209; of leaves, 158; lost, 143; of Lyla June Johnston, 266; of the Māori, 233–34; of mourning, 92; of the Pottawatomi in pre-Chicago, 90; of rivers, 141; by Sean Prentiss, 32; for seeds, 225–31, 270; and slavery, 52; traveling through time, 198–99; to water, 79–80, 83; of the Winnemem Wintu, 83
Sonoran Desert bioregion, 136
sorghum, 150
South Dakota, 142
Southern Gulf Islands (Canada), 177
Spiller, Chellie, 232–39, 269–70
spirals, 157, 161–62, 164, 252
spiritual amnesia, 250
spiritual archetypes, 135
spiritual centeredness, 112
spiritual experience, 247–48
spiritual healing, 208
spirituality: American Indian, 170; of corn, 210–12; evolving, 135; as interconnectedness, 190; as iwígara (interconnectedness of all things), 168; Māori, 232–33, 238; misguided, 80; of Mount Shasta, 80, 127–28; and physical materiality, 190–91; as place-connected self-identity connected to place, 170; Rarámuri, 168; recovery of, 250; as shared experience, 170; and technology, 104; of the Unangan people, 242
spiritual practices, 187

spiritual resilience, 99, 190
spiritual sickness, 248
spiritual teachers, 4
Split-Head Society, 220, 261n1
Spring Creek Prairie (Nebraska), 23
springs, 73, 79, 81–83, 172, 177
Squire, Louise, 179
Standing Rock Sioux Tribe and Reservation, 77–78, 143–44, 260n4
stars, 149–50
St. George Island (Alaska), 243, 245
Stonehenge, 61
Stories from the Leopold Shack (Leopold), 139
St. Paul Island (Alaska), 242–43, 245
Streit Krug, Aubrey, 20–24, 270
suicide, 54–55, 188, 190, 245
sun, 64, 146–47, 150–51, 153, 203
Sweeney, Manea, 87, 95–99, 270
Swentzell, Rina, 133
Swimme, Brian, 150
Systems View of Life, A (Capra and Luisi), 131

Tanana Dene Alaskans, 108
Tao (oriental life-force philosophy), 160–61, 162, 165
Taos Peace and Reconciliation Council (New Mexico), 266
Tao Te Ching (Lao Tzu), 159–60
Tawhai, Fraser Puroku, 235
terminator seeds, 189
Tewa Pueblo Native Americans, 133, 270
Think (family of podcasts), 126
Think Radio studios, 126, 128
Three Sisters, 202, 228
Thunder Valley Community (South Dakota), 142
time: ancestors present in stream of, 50; as a barrier, 18; of the Big Bang, 153; corn feeding the sacred hungers of, 229; cosmological, 212; ecological crises in, 124, 216; generational, 3; geologic time, 139–40, 160–61; as hope, 107; human life challenged in, 127; limitless, 19; in the Māori underworld, 96;

nonlinear, 206; as present time, 39, 126, 250; space-time, 148, 188; timelessness, 216, 226; time-travelling, 234; transcended, 234; of the universal common ancestor, 130. *See also* geologic epochs/periods; "Seventh Fire"
"Time Traveler" (Johnston poem), 197–98
tobacco, 33–34, 39, 223, 226–27, 263
"To Future Kin" (Calvert poem), 89
Tohono O'odham Nation, 136, 271
"To Hope of Becoming Ancestors" (Johnson and Warren essay), 100–109
Tonga Polynesian Kingdom, 220
Tongva Indigenous people of California, 214
"To the Children of the 21st Century" (Kakugawa poem), 44–45
Triassic period, 161
Trudell, John, 3, 220
Trump, Donald, 63, 106
Tucker, Mary Evelyn, 150
Tule grasses, 76
Turtle Clan (Seneca), 264
Turtle Island, 3–4, 52, 54, 210, 220, 264
Turtle Mountains (Central North America), 219, 224, 269
Tuskegee University, 56–58, 267
Tuwaletstiwa, Philip, 133–34

Uintah Reservation, 260n5
Ukraine, 142
Umóⁿhoⁿ (Omaha) Tribe, 21–22, 24
Umóⁿhoⁿ íe tʰe (Omaha language), 21
Unalaska Island (Alaska), 246–47
Unangan (Aleutian) Islands and people, 240–50
Unsettling of America, The (Berry), 35, 263
"Unsigned Letter to a Human in the 21st Century" (May poem), 6–9

van der Kolk, Bessel, 93–94
Van Horn, Gavin, 159–65, 270
Vermont, 25–27, 125, 264, 269
Vernon (California), 214

Wagenberg, Nicola, 200–209, 270
Wakontah (Osage Great Mystery concept), 162
Wald, George, 149
Warren, Julianne Lutz, 100–109, 145–53, 270
warriors, 142, 197, 199, 234
Wartes, Alan, 126–27
Washington (state), 139
water: clean, 229; cognizance of, 74, 76–77; cultural advances in regions near, 152–53; cultural traditions sustained by, 207; guardians/protectors of, 121; and the human body, 160; industrial farms' usage of, 82; life sustained by, 207; as model for human life, 159; oxygen depletion in, 77–78; as pathway, 122; powers of, 159–60, 163; and romantic love, 162–63; as sacred, 81, 83, 96, 236; shortages, 70, 77, 182; sprayed ritually, 167; in the *Tao Te Ching*, 159; unhealthy, 82; unjust distribution systems of, 125. *See also* ancestral waters; GOTW (Native Youth Guardians of the Waters Program) (San Francisco Bay Area)
water pollution: by acid rain, 128; since the 1800s, 77; mothers exposed to, 106; in New Zealand, 98; reversal of, 145, 150; in Salmon runs, 79, 82–83; and unfulfilled promises, 84; visions of, 98
Water Wars (Shiva), 189
Wayfinding Leadership (Wolfgramm, Barclay-Kerr, and Panoho), 236, 270
Weaver, John, 22
Webb, Sky Road, 1
Weidman lineage, 28
Wendell Berry Farming Program (Kentucky), 263
Western Colorado University, 125–26, 255, 265
"What Is Your Rice?" (Hausdoerffer essay), 124–29
"What Kind of Ancestor Do You Want to Be? A Plantcestor!" (Ferrer song), 202–3
"What Kind of Ancestor Will I Be?" (Avalos poem), 201–2

"What Passes, What Remains" (Berry), 35
White, Rowen, 225–31, 270
White Earth Reservation (Minnesota), 1, 5, 125–27, 142, 187, 190
white people, 65–66, 103, 261n1. *See also* light-skinned people
white supremacy, 65–66, 185
Whyte, Kyle, 23–24
Wildness (Hausdoerffer), 125, 127, 265
wildness/wild places, 128, 146–48, 159, 174
Williams, Brooke, 61–67, 271
Williams, Crystal, 110–16, 271
Winnemem Wintu Tribe: Big Head spirit beings among, 79, 83; bounties on, 76; Creator of, 73–75, 77–79; fasts of, 74; Florence Jones as healer among, 268; lands threatened, 215; of the Mt. Shasta/McCloud River watershed, 260n6; Run4Salmon ceremony of, 75–76; salmon in the culture of, 73–76, 78–79, 82–83; salmon restoration by, 209; vs the Shasta Dam, 75–76, 82; Sisk as Tribal Chief of, 4, 269
Wisconsin, 138, 267–68
Wisconsin River, 138, 141
Wisdom Keeper (Merculieff), 240, 249–50
Wise, Eryn, 47

women: abused, 111; in African American families, 110; all-Indigenous women's trust, 261n15; ancestors of the Nelson lineage, 219; books about, 260–65; ceremonies conducted by and for, 250; displaced during menstruation, 188; enslaved, 173; Indigenous women as victims, 53; infertilization of, in Sub-Saharan Africa, 60; National Women's Hall of fame, 267; as Ohlone leaders, 209; patriarchal disrespect for, 66, 192; persecuted as witches, 192; as planters, 188; of power, 269; productions of. disregarded, 192; shelters for, 7, 250; Umó^{n}ho^n Women Elders continuing Native traditions, 21; with or without children, 222; young and old working together, 249
Wyoming, 140

Yama (cosmic artist), 206
"Yes, I Will" (Kakugawa poem), 117
"Your Inheritance" (Kakugawa poem), 194
Yule Pass (Colorado), 125–26, 128
Yup'ik Nation (Alaska), 201

Zuni Tribe of the Zuni Pueblo, 212, 260n5